老屋裝修基礎課

東販編輯部——編著

CONTENTS

CHAPTER 1 關於老屋，你該懂的這些事

Point 1　從專家思維了解老屋裝修

- 011　・日作空間設計
- 014　・賀澤設計
- 018　・日居設計
- 022　・隹設計

Point 2　買屋前該知道的事

- 027　01｜挑選老屋
- 032　02｜翻修費用
- 035　03｜相關法規

Q&A
- 037　Q1. 想買老屋翻修，可以找設計師一起看房子嗎？
- 038　Q2. 聽說老屋翻修費用比新成屋高出許多？
- 039　Q3. 老屋翻新後，大約還能住幾年？
- 040　Q4. 整棟公寓大家陽台都外推，我也能外推嗎？
- 041　Q5. 和新成屋相比，老屋翻修有什麼優勢？
- 042　Q6. 現在住的老屋約有十幾二十年，我的房子是否該做翻修了？
- 043　Q7. 在進行老屋翻修前，有什麼要注意的事項？
- 044　Q8. 我家算是老屋嗎？屋況檢查，要檢查哪些地方？
- 044　Q9. 如果只是局部翻修，也要申請裝修申請？
- 045　Q10. 買老屋好貸款嗎？

CHAPTER 2 老屋裝修，可以這樣做

Point 1　老屋工程這樣做

- 049　01 ｜ 基礎工程
- 058　02 ｜ 裝修工程

Q&A

- 064　Q1. 明明基礎工程全部打掉重做，但還是出現漏水問題，是工程沒做好嗎？
- 064　Q2. 裝修老屋時，哪些舊有元素可以保留使用
- 065　Q3. 房子二十幾年了，看起來問題不大，一定要花大錢翻修嗎？
- 066　Q4. 所謂的基礎工程是指哪些？
- 067　Q5. 牆面出現壁癌，可以直接批土油漆處理掉嗎？
- 067　Q6. 舊家具應該要保留嗎？
- 068　Q7. 因為拆牆留下一根柱子，怎麼設計才能看起來不奇怪？
- 069　Q8. 正要進行老屋翻修，地板看起來好像狀況不錯，一定要拆掉重做嗎？還是有什麼其它方式可以處理？
- 070　Q9. 老屋哪些一定要做？哪些可以不做？
- 071　Q10. 老屋翻新流程和新成屋一樣嗎？主要有哪些流程？

Point 2　破解老屋格局問題

- 073　01 ｜ 缺少採光
- 076　02 ｜ 動線不順
- 079　03 ｜ 空間不方正
- 081　實例示範

CHAPTER 3 空間實例

088	Case01	拆牆重塑動線讓生活變流暢
092	Case02	釋放封閉廚房，讓長形老公寓滿溢日光
096	Case03	大幅挪移客浴，打造寬闊無障礙退休宅
100	Case04	旋轉美梯煥新 40 老屋陳舊感
104	Case05	打好老屋基底，迎接未來各種生活可能性
108	Case06	重整結構與漏水、機能升級，50 年老屋迎來新風貌
112	Case07	微整 33 坪老屋，廚衛調度更靈活
116	Case08	舊宅重生，機能與美感並存
120	Case09	挪一房換書牆大廳變寬變美
124	Case10	融入藤編、海棠花玻璃與綠色，營造自然沉穩復古風
128	Case11	少一房收整廊道，重塑老公寓的當代風貌
132	Case12	拆臥室、不做客廳，50 年老屋放大兩倍空間感
136	Case13	結合天然材質與無障礙設計，不只活化老宅住得更安心
140	Case14	甜漾木白‧和菓子居家
144	Case15	挪移主臥和衛浴，打造雙主臥格局

148	**Case16**	洄遊於前庭後院的簡約日宅
152	**Case17**	10 坪老屋全開放，重塑光感美學宅
156	**Case18**	老屋格局重整變開闊，收納加倍升級
160	**Case19**	30 年老公寓蛻變溫暖北歐風
164	**Case20**	四房變五房，打造三代共享的無障礙退休宅
168	**Case21**	30 年陰暗老屋，變身開闊淨白大宅

172　DESIGNER DATA

CHAPTER

1

關於老屋，你該懂的這些事

老屋裝修基礎課　　010

Point 1 從專家思維了解老屋裝修

空間設計暨圖片提供｜日作空間設計　文｜Fran Cheng

日作空間設計總監

黃世光

老屋除了虛坪少、面積足,多數老屋設備已陳舊,就算全部拆除也不可惜,加上格局、機能能根據新主人需求重新調整,讓生活氛圍更有機會客製化,這應該是老屋最迷人之處。

老屋讓客製化裝修更可能實現

　　新屋房價直直漲讓人買不下手,不少屋主轉而決定買老房子,希望能在相對較親民的老屋市場尋寶。然而,老屋裝修除了要擔心可能因屋況未明而踩雷外,老屋安全性、格局合理性或機能是否真能新穎化?每每成為屋主心中隱憂,裝修繁雜程度也遠比新成屋高出許多,讓屋主深感忐忑。針對這些疑惑,就聽聽多次經手老屋翻新的日作空間設計總監黃世光來為您一一解答。

翻新老屋值不值得?關鍵是經費與時間

　　除了擔心踩雷,屋主更想問:買老屋來作翻新設計到底值不值得呢?總監黃世光認為:「關鍵在於經費與時間。只要屋主時間不趕、願意等,經費也足以因應支撐,老屋裝修的結果都會是好的。」看看這些少說二、三十年,甚至四、五十年的老屋,因為早年公設比較低,屋內實坪足,可使用空間更大;加上基地多位居於優質生活圈或學區附近,讓老屋生活的方便性加分不

◉位於一樓的老屋不僅能有闊綽的庭院，窗外綠帶，透過開窗調整就能改善採光，座擁完美窗景。◉拆除原本不合宜的隔間，將格局依據屋主自己的需求重新調整，讓動線不再制式化，生活軌跡更自如隨興。

少。黃世光接著說明：「一般新房子設備、格局都是新的，買後若有不合意之處想拆換、想改建都會覺得可惜，屋主多半就將就著住了。但是老屋年久老化的設備打掉也不覺可惜，其它如門窗、隔間牆、格局的可更動性也比新屋更高，讓空間可根據新需求來重新調整。」

老屋裝修比起新屋每坪粗估需多 5 萬以上

老屋翻新與新成屋相比，雖然顯見有更多自由度及更高客製化程度。不過，黃世光也坦言：「老屋裝修確實必須付出更多金錢支出，由於每一間老屋的屋況都是獨一的，每位屋主的要求也不盡相同，所以預算支出會比新成屋更難估算、落差也更大。只能試著以粗略的工程預算來說，如果新成屋裝修一坪需要 10～12 萬，老屋每一坪可能會需要 15 萬起跳，所以一坪大約需多準備 5 萬元，才足以支應屋況修復或格局重整等費用。」

但是因為老屋的虛坪少、面積足，加上各種在地情感串連等優勢，讓老屋翻修也成為不少屋主首選標的；此外，有些父母傳承的老屋甚至是金錢無法估算的價值，也讓許多屋主即使花上二至三倍的時間成本，或付出更多經費都願意為老屋做翻新。

在舊有基礎上創新的舒適空間

屋主在老屋翻新時通常最關注的是屋況修復，這當然是舒適新居的必要基礎，尤其修復過程除了繁瑣，也可能狀況百出，同時各種變更還需要經過管委會同意，或與公寓使用執照變更有關，牽動的層面相當廣泛，也足見設計團隊的經驗值很重要。

除此之外，老屋翻新的另一個特色是故事性，在合作過程總是抱持理解與包容，與客戶耐心溝通的黃世光，以過往裝修經驗來解說：「有些屋主會希望保留舊家具，所以如何將這些有年代感但還堪用，或是與屋主有密切連結的物件融入新空間，也是老屋裝修重點，甚至採用老屋原結構上的優美元素，在這些既有基礎上創造出新穎舒適的空間，也能延續空間故事性，這是新成屋無法達成的裝修效果。」

老屋裝修基礎課

先了解老宅體質，
再打造安心居住宅

空間設計暨圖片提供｜賀澤設計　文｜陳佳歆

賀澤室內設計總監

張益勝

> 老屋改裝以居住安全及舒適度為主，找一位對老屋熟悉的設計師事先瞭解屋況，重新調整好房屋的基本體質，並釐清居住者實際使用需求和目的，給予合適的空間格局改善昏暗的空間感，才能改造出適合居住者使用的舒適空間。

　　討論老屋翻新的需求跟族群大致有幾個樣貌，一種是購屋預算上希望買到相對坪數大小的房屋，另一種是長輩留下來的老房子，或者是從年輕住到老的自宅，隨著年齡經歷三階段的裝潢。一般來說，屋齡 10～15 年的稱為中古屋，20～30 年的房子就算是老屋，四十年以上則稱為老舊建築，中古屋、老屋和老舊建築翻新都有各自要注意的地方。

留意不同階段的老屋隱藏狀況

　　四十年以上的房子較多是加強磚造，所謂加強磚造是主結構樑柱為鋼筋混泥土，其他外牆跟隔間牆幾乎都是紅磚疊砌而成，在樑柱灌漿前，會讓紅磚事先疊砌出部份企口，咬合在樑柱之間；而中古屋和老屋較多為 RC 結構，一般耐用年限約是 60 到 65 年左右，以耐震度來講，鋼筋混泥房子的耐震度會比磚造來得高，因此都更都是針對四十年以上的房子居多；裝修老舊建築前要注意的是，剪力牆有沒有明顯的破壞、龜裂，類似有海砂屋或是樓板鋼筋裸露等，或樑柱是否有坍塌之類結構的安全問題，賀澤室內設計總監張益勝表示：「這些問題需請專業結構技師來鑑定是不是達到危樓等級，才能進一步進行室內裝修，裝修前也要做結構補強才能確保居住安全。」

通常20～30年的老屋要特別留意漏水問題，像是上下樓層樓板、窗角附近、管道間，還有二次施工的外推位置，這些都是容易發生漏水狀況的地方，因此，假設購買老屋進行修繕，就要針對這幾個地方先做初步檢查。中古屋多是RC結構，安全性比較不會有太大問題，唯一要特別注意的是配電，因為早期房子大多使用直徑1.6mm的電線電纜，而現代居家用電需求比以前高出許多，因此工程配線單線直徑至少要配置2.0mm以上，才能更安心使用家庭電器。

除此之外，早期老屋最常有採光不良問題，這是因為很多老屋會把餐廳規劃在房子中間，使得空間昏昏暗暗的，賀澤室內設計總監張益勝表示：「人生活中需要陽光、空氣、水，老屋翻新同樣注重這幾個元素，其中陽光是最優先考慮，老屋翻新設計中會儘可能改善昏暗的空間感，讓開放領域可以有雙面採光，或者至少有兩面窗戶讓光線可以交叉、互補，提升整體空間明亮度。」

預算有限，先做好打底基礎工程

無論是中古屋或老屋，在原始屋況上都有不少狀況需要解決，因此如何分配預算也是很多人在裝修前會擔心的問題，關於這點，賀澤室內設計總監張益勝建議：「如果是超過二十年以上的老屋，建議先做完基礎工程，相當於把基本體質重新調整，有額外預算再來做裝修。」也就是說，居住安全是最基本也是最重要的事，而其中防水最不能忽略，如果窗戶漏水會影響到自己的居家空間，若是廁所漏水則影響到樓下鄰居，事後修繕可能會衍伸許多麻煩的問題，因此在預算許可下，從居住安全性考量，老屋建議全室水管、電線更新，同時翻新衛浴。

賀澤室內設計總監張益勝也提醒，裝修老屋的重點是人可以住進去，且能舒適及安心的生活，不是要花很多錢把空間做得很漂亮，如果預算全部用完，沒有家具及基本設備一樣沒辦法住得舒服，所以在討論工程過程中，要同時考量到空調、家電、家具、軟裝等預算。

然而,裝潢預算分配也會因中古屋、老屋和老舊建築而有所不同,50～60年的老舊建築建議先做好基礎工程,其他東西盡量精簡,因為可能不久要面臨都更,15～20年左右的房子相對來說結構比較安全,居住時間可能也比較長,可以在裝飾、軟裝上分配較多預算。

老屋通常隱藏許多潛在問題,若想選擇老屋裝修,要找對老屋狀況有經驗的設計師,在裝修前先了解整體屋況,釐清居住者的實際使用需求和目的,從長遠居住思考再著手進行翻修,就能打造出安全舒適的居住空間。

●早期老屋格局餐廳大多規劃在房子中間,老屋改造儘可能改善採光問題,利用前後開窗提升空間明亮感。

老屋裝修基礎課　　　　018

從基礎結構到格局優化，
破解老屋翻修

空間設計暨圖片提供｜日居設計　文｜Celine

日居室內裝修設計團隊

老屋裝修是一項充滿變數的工程,透過專業的規劃與適當的預算控管,可以大幅降低施工風險,並確保最終的居住品質。無論是基礎結構的穩固、電水配管的更新,還是格局動線的優化,都應納入完整考量,以確保翻新後的空間安全又舒適。

掌握基礎工程,確保居住安全

老屋翻修首要關鍵在基礎工程,這部分往往佔據整體預算的大宗,卻是無法妥協的重點。由於舊屋結構與內部狀況難以在購買時全面掌握,拆除後才發現問題的情況相當常見,例如隱藏的管線老化、天花板內部樑柱結構不如預期等。

因此,在預算規劃上,建議屋主額外準備至少 10% 的預備金,以應對不可預測的施工變數。除了基礎結構,電線與水管的更新也是裝修老屋時不可忽視的部分。老舊電線可能無法承受現代家電的高負載,容易產生安全隱患,因此建議全面更換配線,並調整電箱配置以支援現代用電需求。同樣地,水管若有鏽蝕或老化現象,最好一併更新,確保日後使用無虞。

此外,在老屋翻新工程中,電箱與配電系統的更新至關重要。由於舊有電箱的迴路數量有限,許多插座共用同一迴路,容易導致跳電問題,為解決此問題,建議更換為容量更大的電箱,並增加迴路數量。此外,過去的電線線徑可能不足以承載現代電器的高電流,長期滿載可能導致電線過熱甚至熔解,增加火災風險。因此,應加大電線線徑,至少符合現行法規標準。除了

安全性,生活便利性也需考量,老房子插座數量通常不足,難以滿足現代人對3C產品的需求,因此,建議在裝修時與設計師充分溝通,規劃適合自身需求的用電環境,新增插座至理想位置,提升生活品質。

動線優化與空間再造

在空間規劃上,老屋常見的挑戰包括格局狹窄、採光不足及不合宜的隔間配置,針對這些問題,建議拆除不必要的隔間,以創造更流暢的動線,提高空間利用率。例如,狹長型老屋常面臨單面採光限制,此時可運用玻璃隔間或拉門來區隔空間,保持視覺穿透感並提升室內採光。

此外,老屋的結構樑柱可能影響空間佈局,設計時須妥善規劃,例如運用櫃體包覆樑柱,或透過局部微調隔間來改善空間機能,使動線更順暢。家具與收納設計方面,則可先保留必需品,如衣櫃與主要收納櫃,而次要的配件則可視預算逐步添購,避免一次性投入過多費用。

陽台外推的考量與法規限制

在老屋翻修中,許多屋主希望透過將陽台外推來擴大室內空間。然而,這種做法涉及多方面的考量,必須謹慎評估。

結構安全:建築物的結構設計是經過精密計算的,擅自將陽台外推可能破壞原有結構的強度。新增的建材會增加建築負荷,導致受力不均,增加在地震等外力作用下傾斜或倒塌的風險。

消防安全:陽台通常作為緊急情況下的逃生路徑或救援空間。將陽台外推並封閉,可能影響逃生和救援效率,增加危險性。

防水問題:陽台長期暴露在外,容易受到風雨侵蝕,導致漏水和壁癌等問題。將陽台納入室內,可能使這些問題延伸至室內空間,影響居住品質。

法規限制:根據《建築法》第25條,未經主管機關許可,不得擅自變更

建築結構。陽台外推涉及外牆拆除和增加樓地板面積，屬於違法建築，可能面臨罰款或強制拆除。此外，許多公寓大廈為維護結構安全和外觀一致性，也禁止住戶進行陽台外推。因此，在考慮陽台外推時，應充分了解相關法規，評估結構和安全風險，並尋求專業意見，以確保改造合法且安全。

● 老屋翻修預算應著重基礎工程的修復，建議預留10%裝潢預備金，至於軟裝家具可逐步再做添購。

老屋裝修基礎課

老屋裝修挑戰多，
精準掌控施工項目與預算

空間設計暨圖片提供｜隹設計　文｜EVA

佳設計

陳俊宇、彭祐理

> 老屋裝修,最應該優先投入的是基礎工程建設與格局調整,重新梳理動線、採光,同時更換老舊水電管線和設備,才能確保居住安全與舒適度,其餘的收納、軟裝可後續補強。最重要的是別低估老屋的潛在問題,過程中做好檢測與規劃,才能裝得安心,住得放心。

在房價高漲的時代,購買中古屋或老屋成為許多屋主的選擇。與新成屋相比,老屋相對公設比低、空間坪數較大,價格相對有競爭力。然而,購買老屋所面臨的是後續的裝修問題,不僅是風格改造,更涉及房屋基礎結構的檢測與更新,若未妥善規劃,可能會遇到預算超支、施工延誤,甚至影響居住安全。因此,在開始裝修前,建議深入了解老屋翻修關鍵,做好預算與施工計劃,才能打造符合期待的理想居家。

裝修前檢查屋況,漏水與採光是關鍵

佳設計提到,老屋裝修和新成屋的裝修工程差距不大,通常最需要解決的是漏水問題,建議仔細確認天花板、牆角與窗戶接縫處,或是衛浴、廚房濕區相鄰的牆面是否有滲水或壁癌的痕跡。有些漏水問題相當隱密,拆除後才發現是很常見的狀況,有些甚至是鄰居水管破舊造成的漏水更是難以察覺。因此施工過程中,若有發現任何漏水跡象都要及時處理,以免裝修完畢再拆除重做,更耗時費力。

此外,老屋格局體質多半先天不良,容易有樓高低、採光不足的問題,或是房間數量多,導致空間狹窄、動線不順。建議能重新梳理格局動線,拆

◉採光不足和空間狹窄是許多老屋的問題,拆除不必要的隔間,並運用玻璃、拉門或半高櫃體維持視野的通透度。◉適當採用減法策略,降低木作、系統櫃項目,有效節省裝修費用。

除不必要的牆面，改用玻璃拉門、開放式設計，維持明亮通透的環境，避免無效空間的浪費。若是透天厝，則能適時打鑿天井，光線由上而下穿透整個空間，有效提升整體採光。

保留 20％預算，預留突發狀況

近年來，由於人工與材料成本上漲，設計師建議：「目前老屋裝修至少每坪要抓 13 萬元的預算。」而老屋在裝修過程中，經常會發現隱藏的屋況問題，導致需要追加預算，建議需額外保留 20％的預算作為緊急預備金，在使用上能更有餘裕。

若本身預算有限，勢必要把錢花在刀口上，設計師指出：「老屋基礎工程建議不能省，像是水電、泥作防水、窗戶與衛浴該優先更換，房屋體質穩固了，才能提升居住品質與安全。」這是因為水管、電線藏在牆內，管線可能有嚴重鏽蝕或漏水，電線也因長期負載不足，導致跳電或短路。此外，傳統鋁窗的膠條經過長期使用，也容易有脆化問題，不妨趁著裝修一併改善翻新。

其餘裝修項目可視需求增加或減少，降低初期投入的負擔。像是木作、系統櫃採用「減法策略」，盡可能減少施作的項目和範圍，有效節省費用。建議以現成家具來替代，相較於傳統木作和系統櫃，現成家具價格較低，使用也更靈活，未來可依需求更換或增添。此外，若原為磁磚地板且狀況良好，也能直接沿用，或是鋪設超耐磨木地板，而不必拆除舊地磚，不僅能節省拆除與泥作費用，也能縮短施工時間。

最後，違建問題也是老屋裝修常見的挑戰，許多老屋陽台外推或頂樓加蓋屬於違建，雖然增加使用坪數，但若涉及拆除或更新，可能會面臨政府強制執法。設計師提醒：「有些屋主認為現存允許的違建空間能翻修整新，但實際上是只允許修繕並維持現狀，若重新翻修，和原先現存樣貌大不相同，就可能被要求拆除。」因此，在裝修前，應確認建物的合法範圍與違建空間是否需要更新，避免花費大筆資金，因法規問題被迫拆除，徒增損失。

老屋裝修基礎課

Point 2 買屋前該知道的事

空間設計暨圖片提供｜日居設計

01 挑選老屋

精準看屋，
避開問題老屋

　　房價居高不下，為了想買間房，許多購屋者便從新建案轉向購買老屋重新翻修改造，雖說老屋價錢較低、坪數實在，加上有些老屋地理環位置佳，但老屋翻修工程比新屋裝潢複雜許多，更別說可能在拆除之後，才發現隱藏大問題，導致費用無限增加，嚴重的還可能還變成危樓。要怎麼避免踩雷，買到問題老屋？最好的方法就是在買屋前做好功課，才能從看屋開始，就成功避開問題老屋。

老屋建築類型

　　隨著時代與生活的轉變，建築種類也會有所不同，除了依喜歡的建築類型挑選，不同建築有不同的問題要注意，以下是幾種常見老屋建築類型。

｡透天住宅

早期人口沒那麼密集，透天厝是最常見的房屋類型，因此很多透天老厝，尤其是台北，屋齡甚至比屋主年紀還大，很多人以為透天厝任意改變格局，應該不會有結構問題，其實不一定。首先，若曾在頂樓、前後院加蓋，要先確認是否合法，接著再看加蓋出來的建築是屋主自己動手，還是有請專業工班來進行建造，因涉及結構安全，要特別注意。另外，透天厝看屋的另一個重點是樓梯動線，通常單一樓面坪數不大，樓梯會佔去較多空間，若原本動線設計不佳，而導致未來裝修需修改樓梯來牽就格局，有可能要付出更多時間與金錢。

空間設計暨圖片提供｜隹設計

●老屋樓梯位置很重要，位置不好容易導致格局難以規劃，事後若要移位，可能要花費不少錢。

◦ 老式公寓

老公寓最常見陽台外推、頂樓加蓋等問題，除了自己想買的房子不要有違建，若鄰居有過於誇張的違建也盡量避開，以免買到未來有糾紛或者結構已不穩當的建築。最後可前往查看老式公寓的外牆、樓梯間等公共空間，看看外牆是否有明顯剝落、裂縫、水漬等現象，這些區域通常是最能反映出建築確實屋況的地方。

◦ 電梯大廈

電梯大廈除了查看自己要買的房子裡外，也要到公共區確認實際屋況。因集合式住宅內有公共設施，要確認公設內容是否合用？出入大門及停車場有無安全疑慮？大廈內部管理是否良好、正常？社區內環境的維護狀況，以及鄰

居有無佔用公共空間等問題，都是購買大廈住宅時要事先觀察的。

買屋時建議不要只去一次，在白天可去看房子的採光、座向、窗型等屋況，晚上則確認與鄰居的隔音效果、社區是否有怪氣味，如下水孔外溢的臭味、或鄰居廚房排煙對著自家窗戶等生活實況。

屋齡

老屋顧名思義，就是有一定年份的房子，再怎麼堅固的房子一定會因為時間變老變舊，而不同屋齡的房子老化程度也不一，各自要面對的問題也不同，因此購買前屋齡也要列入評估。

○ 20～29 年

最資淺老屋，正常情況下建築結構應該堅固，外觀若保養得宜也不會太差，不過，二十幾年前蓋的房子，當時住宅電器化程度較低，用電形式與用電量都與現今不同；加上水電管路均為耗材，原則上 15～20 年就應換新，因此，即使購屋後格局適用不需重新裝修，也可能要重新規劃水電管路。

○ 30～39 年

屋況多半已呈現自然老化，水電管路應該要全面更新，牆面可能出現壁癌、漏水現象，所以少不了要進行鋁窗、泥作、木工方面的工程。此屋齡房子還有一件事要特別確認，由於三十幾年的房子正值民國七十幾年的海砂屋、輻射屋等汙染嚴重時期，購屋前應先請仲介提出安全證明；或者上各縣市政府網站查詢，確認是否為海砂屋、輻射屋、地震屋、山坡地房屋等。

○ 40 年以上

四十年以上的老屋除了可從外觀目視有無牆面歪斜，或可用乒乓球測試地板傾斜狀況外，還要弄清楚房子是 RC 結構還是磚造屋，由於此屋齡的房子除了有三十年以上老屋問題外，可能還需以植筋或鋼構方式加強結構安全。另外基本上四十年左右的房子銀行貸款的成數相當低，若位於市中心的話還有土地價值，否則未來脫手可能更難。

屋況

　　雖然都是老屋，根據年份不同，老屋的屋況不盡相同，但實際屋況對於購屋後翻修的費用影響極大，因此若能在一開始看屋時就對屋況做出預判，多少能避開問題老屋。

○ 管線老化

主要鎖定電路與水路，專家建議超過十五年就需考慮將管線全面換新。尤其電路不易檢查、又具危險性，加上現代家庭電器用量大幅增加，而早期的總電量安培數多半已不敷使用，必須重新向電力公司申請。

至於水路，除了依漏水情形來決定要不要重新佈線外，早期住宅水管的材質若為鐵管、塑膠管則有生鏽、卡垢疑慮，產生飲用水健康問題，建議還是換掉。而除了室內水管，也應查看幹管是否有漏水問題，還有就是由外面接進來公共幹管若未更換，即使室內已經換管，但從水塔至居家段仍有汙染問題，因此需從公共水塔直接拉管重設。

○ 設備老舊

設備包括衛浴間內設備、廚房三機、冷氣空調等，是否需要更換的考量重點依序為：1. 設備是否還堪用？ 2. 機能是否合用？ 3. 衛生性及舒適度屋主是否能接受。通常若有重新裝修設計者設備多會跟著換新，但有些老屋可能近幾年已做過維修及設備更換，因此，新屋主可依照自己的需求來決定是否沿用，若預算不足也可日後再作更換。

○ 結構損壞

老屋最擔心前屋主們是否曾作過結構變更，尤其若未經評估就變更格局可能造成房屋結構失衡，導致危險，應請前屋主提供房屋「竣工圖」，若有樑柱變更過盡量避開不買，或另請結構技師作評估。若室內已重新裝修過無法觀察屋況，可由公共設施如外牆、Lobby、樓梯間等地方，詳細查看是否出現每層

樓同方向性的裂痕、頂板混凝土塊剝落以及樑的水平裂縫、柱子的垂直裂縫，藉此評估屋況安全性。

◦ **漏水**

此部分漏水主要指牆面、天花板的問題，尤其在外牆部分最簡易的勘查就是察看有無明顯水痕；但室內若已有裝修過恐怕不容易看到，建議可以從公設去看，了解外牆或屋頂防水是否需重作。

◦ **蟲害**

許多老屋有蟲害問題，主要是室內潮濕造成寄生的蟲卵得以孳生繁衍，需要依個別屋況檢視，了解發生蟲害問題是在於氣候、環境，屋內通風不佳、濕氣不易蒸散，或是木作家具導致，若屋內有木結構者更要小心評估，以免購屋後要花大錢作結構修復，恐怕得不償失。

02 翻修費用

費用精打細算，
老屋裝修預算不爆表

老屋翻修除了最重要的施工工程外，最讓屋主煩心的就是錢的問題，裝潢費用應該要怎麼編列、分配運用？老屋翻新又能向政府申請什麼補貼？這些都是所有屋主最關心的事，以下整理出一些相關訊息，提供屋主一些參考。

老屋補貼

全台老屋數量不少，為了幫助民眾住的安全、安心，其實政府提供許多老屋相關費用補助措施，而且除了中央政府補助優惠方案，各縣市也有相關補助條款，有需求的人可到相關網站查詢。

01 ｜內政部修繕住宅貸款利息補貼

由內政部推出的住宅修繕貸款利息補貼，符合條件並獲得補助的民眾可持補貼證明向金融機構申請優惠修繕貸款。

申請方式：

書面申請	採書面申請方式者，申請人於受理期間，填寫申請書並備妥規定的應檢附文件後，以掛號郵寄至申請人戶籍所在地的直轄市、縣（市）政府【詳直轄市、縣（市）政府受理單位】，申請日之認定以郵戳為憑。
線上申請	申請人於受理期間，至內政部建置住宅補貼線上申請作業網站提出申請並上傳相關文件。

申請人資格、條件：

基本符合	中華民國國民在國內設有戶籍，且符合下列規定之一 ● 已成年。 ● 未成年已結婚。 ● 未成年，已於安置教養機構或寄養家庭結束安置無法返家。
補充說明	● 家庭成員住宅持有說明 家庭成員僅持有一戶住宅且其使用執照核發日期逾十年，該住宅應為申請人所有、其配偶所有或其與配偶、同戶籍直系親屬、配偶戶籍內直系親屬共同持有。 ● 申請時家庭成員均未接受政府其他住宅補貼。 ● 家庭年所得及財產應符合住宅補貼對象一定所得及財產。

※ 相關應檢附內容，可上內政部不動產資訊平台詳閱。

02 ｜老舊建築物修繕補助

符合申請條件之合法建築物，可申請以下補助：
1. 外牆安全整新：最高 400 萬元。
2. 結構安全補強：最高 500 萬元。
※ 上述補助不超過總工程經費 50%，但若屬市府公告劃定之整建住宅（以下簡稱整宅）或適用都市計畫劃定山坡地開發建築管制規定地區（以下簡稱山限區），補助金額上限更高。

03 ｜老舊建築物增設電梯補助

符合申請條件之合法建築物，每棟建築物補助 1 座電梯，以 300 萬元為上限，且不超過總工程經費 50%。符合特定條件者（華江地區重建範圍住宅、山限區、整宅等），補助金額上限更高。
資料來源：內政部不動產資訊平台、我的 E 政府

裝潢費用分配運用

與新成屋不同,老屋在翻修過程中,很容易發生意想不到的突發狀況,為了應付更多不可預期的施工變數,裝潢費用要怎麼編列與分配運用,才能在有限的預算裡,將老屋翻修成理想中的居家空間,同時又住得安心且安全呢?

比起新成屋,可以預期老屋拆除後會應該會遇到更多問題要解決,因此若是一般裝潢費用一坪預估約 7 ～ 10 萬,老屋則可能要提高到 12 ～ 15 萬,才能應付突發狀況,至於預備金,則建議要預留總工程款的 10 ～ 20%,使用上才會更有餘裕。

裝潢工程的進行,新成屋、老屋其實差不多,差別在於老屋不只房子表面問題要修繕,更多的是拆除之後的隱藏屋況要處理,也因此裝潢費用運用比例,會和新成屋有所不同。一般來說,老屋基礎工程佔據費用最大宗,也就是大約 30 ～ 50%左右,會依實際屋況再來做增加或減少,確定了基礎工程比例之後,剩下的預算才會再依需求分配給木作、系統家具等其它工程。

雖說老屋打好基礎很重要,但若沒有家具、軟裝、空調等,真的住進去也不會感到舒適,在討論預算時,記得最好留有預算,來添購家具、軟裝。若整個預算用起來還是捉襟見肘,那麼可以從木作和系統家具這兩項工程來做節約,盡量減少木作施作與施作範圍,或是以現成家具做取代,應該都能省下不少。

空間設計暨圖片提供｜隹設計

●除了基礎工程預算不能省,其餘像是木作和系統工程,是最好精減的項目,可以減少施作木作工程範圍,或以現成家具取代系統家具。

03 相關法規

仔細了解法規，
小心犯法被罰款

進行老屋翻修時，常常為了調整格局，而需要拆除牆面或樑柱，又或者原本就有外推陽台，想再次翻新或擴建，這些老屋翻修時常見的狀況，可能會影響房屋結構安全，又或者會與現有法規有抵觸，在翻修時都要特別注意。

1. 承重牆與樑柱不能隨便拆除

老屋常見不符合使用的生活格局，為了要調整格局，可能會需要拆除部分牆面或樑柱，要特別注意的是，承重牆絕對不可隨意拆除，因為一旦移除可能破壞建築整體穩定性，通常遇到這類情形，需請專業的結構技師或結構技師事務所評估處理。

2. 陽台外推小心違法

過去陽台外推的情況普遍嚴重，因此老屋翻修最常遇到有陽台外推，由於陽台外推本來就屬違章建築，因此不只陽台不能外推，若購買的房屋已有外推情況，翻修時不能再做翻新更動，最好在申請室內裝修許可時，附上屋況資料，證明該違建並非自行改動。

3. 頂樓加蓋可能被拆

雖然現在許多老公寓頂樓都有加蓋，部分縣市政府也會將老舊違建列為暫緩拆除，但其實頂樓加蓋為指屋頂平台突出物增建，凡是在頂樓建築面積 1/8 法定範圍內屬合法建物，超過此範圍即違法，因此即便只是單純想在頂樓加蓋雨棚遮陰，都可能算是違法且即報即拆。

4. 室內裝修要申請許

有些人可能以為只是自住的老屋要自己發包翻修，不用申請裝修許可，其實依據《建築物室內裝修管理辦法》第3條規定，室內裝修除壁紙、窗簾、家具等外，只要涉及天花板、牆面裝修，及超過1.2公尺高度的固定隔屏、分間牆的變更等，都屬於需申請許可範圍。

裝修住宅只要是六層以上的集合住宅（有3個住宅單位以上之建築物）或五樓以下（含五樓）的公寓或大樓，有更改隔間或變動格局的狀況就需申請。若翻新的是自有透天厝可免申請，但五樓以下集合住宅若涉及「新增廁所」、「新增2間以上居室」等一樣需要申請許可。

裝修前要完成申請許可手續。若未經合法審查擅自裝修，屋主、使用人或業者將依法罰鍰，並限期改善或補辦，逾期不改者，得連續處罰或強制拆除。

Q & A

Q1

想買老屋翻修，可以找設計師一起看房子嗎？

在高房價的現在，很多人轉而投入房價較低的老屋做翻修，但要怎麼避免被不可預期的隱藏屋況拖垮，一開始挑對老屋很重要。

比起完全不懂的買房菜鳥，可以找一個有老屋翻修經驗的設計師陪同看房，設計師可以根據過往翻修經驗，從現有屋況來做出初步判定，已有的漏水、壁癌等問題是否值得修復？現有格局是否要做大幅度更動？缺少採光問題能否解決？有了設計師的經驗加持，屋主就能比較明確判斷這棟老屋是否值得買下來進行翻修。

設計師陪同需要費用嗎？要的，由於目前老屋翻修案例越來越多，現在很多設計公司皆設有陪同看屋的這項服務，有需求的人可上網查詢相關資訊，與設計公司收費標準，服務內容通常要視各設計公司規定，每家不盡相同，可多找幾家做比較，這種服務大多都採單一次服務收費，未來不一定要由這間公司來負責後續翻修工程。

空間設計暨圖片提供｜季沃設計

● 老屋翻修難度高，可事先請設計師陪同看屋，比較能避開問題老屋，也較能買到適合自己的老屋。

Q2 聽說老屋翻修費用比新成屋高出許多？

所謂的老屋通常是指二十年以上的房子，像這樣的房子即便屋況良好，屋內格局不用大幅改動，但有些該做的還是不能省，像是水電管線的更新，因為水電管線用久了會開始有損壞狀況，因此一般只要是15～20年的房子，就建議要全屋換新。

再會保養，風吹雨打了幾十年的房子總是會壞會舊，其中最可能出現的就是漏水、壁癌、窗戶不堪使用等問題，這些問題也大多需要藉由泥作、防水工程來處理解決。以上這些都是新成屋裝潢不會有的問題，卻是老屋一定要做的修繕，且只有解決了老舊屋況，才能開始進行一般新成屋的裝修環節。

除此之外，新成屋的裝修通常不會大動格局，所以裝修施工時間不會太長，約2～4個月就能完工，而老屋翻新除了打除舊裝潢，還要進行基礎工程完善結構，接著才能開始裝修，花費時間需要更久，通常約要3～9個月才能完工。

整體來看，老屋要付出的時間與要施作的工程，都比新成屋要多，也因此費用相對會高出新成屋許多，因此建議想進行老屋翻修的人，除了裝潢費用要高估，預備金也要多準備一點，以應付老屋不可預期的變數。

空間設計暨圖片提供｜樂湉設計

● 老屋需將因建築老舊而損毀的屋況修復，除了基礎工程，翻修時間長，因此費用自然會升高。

Q3

老屋翻新後,大約還能住幾年?

公家機關辦公房舍如為鋼骨、鋼筋混凝土構造,使用年限為 60 年,如為住宅用則為 55 年,加強磚辦公房舍為 35 年,如為磚石牆載重者為 30 年。也就是說,從相關法規來看,屋齡 50 年以內的房子仍然在適合居住的範圍內,但是否真的適合再住下去,其實還是要看實際居住年限與施工品質和建築品質。

想要老屋翻新然後住得長久,首先在購屋前就要先留意房子耐震係數,及建商是否提供建築安全履歷,並檢視外觀狀況、評估結構安全,只要基本屋況確定沒問題了,接著就是後續的翻修工程要做確實,將該解決的漏水、結構等問題都解決了,等到入住之後,固定並持續做房屋維修,那麼其實老屋要再住個二十年也不成問題。

空間設計暨圖片提供│佳設計

● 想要翻修老屋住更久,除了一開始就要選擇屋況良好的老屋,做好後續的翻修工程及維修,也能延長老屋居住年限。

Q4 整棟公寓大家陽台都外推，我也能外推嗎？

一般來說，陽台外推屬於不合法違建，且對居住安全有高風險疑慮，不論新舊房子（中古屋或新成屋）都一樣，差別只在有無被查報。若是購買的老屋已有陽台外推，那麼首先要確認是否有被查報，若沒有，可在申請室內裝修許可時檢附上照片與平面圖，證明非自行外推，便可列為緩拆，若是曾被查報，那麼應該要恢復原狀，以免日後被舉報，可能就要面臨強制拆除。

會想要將陽台外推，無非是希望可以獲得更多使用空間，然而陽台承重不如室內空間，若做為生活空間使用，反而會有安全上的疑慮，想讓空間變大，最重要的就是盡量減少過於零碎的隔間，並放大採光條件，如此一來就能讓空間變得比較寬敞，而不外推陽台又想讓空間變得大一些，建議採用可完全收起來的玻璃折門，門片完全收起時便可虛化內外界線，營造出空間延伸效果，玻璃材質則有助加強採光，採光變好放大效果自然加乘，住起來也更舒適。

空間設計暨圖片提供｜日居設計

●陽台外推本就屬於違法，且承重不足，會讓居住安全產生疑慮，若納入室內還容易將戶外漏水問題引入生活空間。

Q5

和新成屋相比，老屋翻修有什麼優勢？

雖說老屋價格上比新成屋低，但不只施工程度複雜，而且施工期還比較長，這樣看來，買老屋翻修真的比較有優勢嗎？其實還是有的，雖說翻修施工時間和花費比新成屋來得高，但公設比卻比新成屋低，也就是說可使用坪數比較大，而且大部分老屋位在市中心，地理環境佳，生活機能完整，而因為市中心可做為建地的土地已經不多，所以新成屋多半位於郊區，生活機能還有待發展。

其實買新成屋或買老屋，應該要看個人需求與條件，實在沒有足夠的資金，又想要有間房，轉而買老屋翻修是一種選擇，但最終還是應從自身思考，是否有足夠的時間等待，又或者對居住空間與未來生活的要求為何，多從幾個面相思考，才能真正找到一個理想中的家。

	老屋	新成屋
公設比	約 10%	約 30～35%
地段位置	老屋大多位於市中心，不只交通便利、生活機也比較完善。	大多位於市中心周圍的郊區，生活機能比較不方便。
增值潛力	由於原本地段就好，經過翻修後，增值潛力高。	購買地段是否增值，易受周邊設施、景氣等因素影響。
施工時間	約需 6～9 個月，甚至可能長達一年。	大約 3 個月左右，即可完工入住。

Q6

現在住的老屋約有十幾二十年，我的房子是否該做翻修了？

房子會老會舊，長年累月下來甚至可能已出現損壞，但已經住了幾十年，很多人可能沒有意識到房子需要翻修，因此若家中已出現以下幾種狀況，會建議可以開始規劃做翻修。

。家庭成員變多
當家庭成員變多，不只每個人使用空間越來越少，各自的生活用品會讓家中變得凌亂，擁擠的空間也導致生活品質變差，這時建議可以進行老屋翻新規劃，藉由翻修工程，可依照現有成員與生活模型來重新調整格局與動線，空間可以更好被利用，也能讓家人生活得更舒適。

。住房修繕與安全結構加強
房子會隨時間損耗，結構逐漸變得脆弱，若房屋已經可見明顯龜裂，或屋齡已超過二十年，這時候就應該規劃翻修工程，藉此不只可修繕已損耗的地方，同時也能適時補強屋體安全結構，確保居家安全。

空間設計暨圖片提供｜森叄設計

●透過老屋翻修，才能量身訂製，打造出符合現有生活模式的空間，不只住得舒適，也有助增進家人感情。

Q7

在進行老屋翻修前，有什麼要注意的事項？

不同於新成屋單純的裝潢流程，面對施工複雜許多的老屋翻修時，有些事項應該要在事前就先注意，以免多走了冤枉路。

。評估實際屋況
在進行翻新之前，應該要找專家，做一個全面的屋況檢測，檢測內容包含有房屋結構、水電管線，及牆體是否有潮濕發霉、漏水壁癌等問題。

。預算分配
由於老屋狀況多，想要在未來住得舒適，勢必要將房子現有問題解決，也因此費用最大比例會落在基礎工程，雖說基礎工程看不到，但能確保居家安全，必須認知基礎工程為重，其它像是家具、軟裝等，可日後再慢慢添購。

。了解相關法規
老屋翻修最常遇到有拆牆、拆樑柱等情形，在進行這些拆除前，要先確認是否符合台灣建築法規，像是承重牆就絕對不能拆，若有涉及違法，除了罰款，還可能影響結構安全。另外，頂樓加蓋的面積若超過建築物8分之1則屬違建，且公寓大廈頂樓所有權屬於全體住戶，在申請裝修許可時要取得所有住戶同意，否則很容易遭檢舉拆除。

空間設計暨圖片提供｜樂湉設計

● 老屋翻修狀況比較複雜，若能事前多做功課，就能減少不可預期的變數發生，自然也能更精準掌控預算和工期。

Q8 我家算是老屋嗎？屋況檢查，要檢查哪些地方？

雖說沒有明文規定，但一般 20～30 年以上即可視為老屋，若屋況不差，基本上還算適合居住，但會建議可以開始規劃進行翻修，而翻修前可先進行屋況檢查，應檢查的地方如下：

。房屋結構
和房屋結構息息相關的有承重牆與樑柱，首先先檢查牆面與樑柱是否有出現龜裂或變形，若有問題，可請專家做評估，進一步制訂補強計畫。

。水管電線
一般只要超過十五年以上的老屋，大多會建議水電全部更新，因為電線壽命約為十五年，而水管則是 10～20 年，而且經過幾十年，過去用電習慣與現今不同，更新電線會比較安全，檢查重點一般是查看電線是否老舊，水管是否有出現堵塞漏水情形，若有則應該更換。

。漏水、壁癌
檢查是否有漏水、壁癌，從牆面來看，可查看是否有剝落、發霉等問題，再來是窗戶是否有確實密實，地面有否破損、潮濕狀況，若是位於頂樓，則還要檢查頂樓屋頂防水是否完好。

Q9 如果只是局部翻修，也要申請裝修申請？

只要進行室內裝修，不論範圍多大，都應該要提出室內裝修申請，以下列出較為明確的申請規定。除壁紙、壁布、窗簾、家具、活動隔屏、地氈等之黏貼及擺設外之下列行為：

1. 固著於建築物構造體之天花板裝修。
2. 內部牆面裝修。
3. 高度超過地板面以上一點二公尺固定之隔屏或兼作櫥櫃使用之隔屏裝修。
4. 分間牆變更。
涉及上述四點，就必須申請室內裝修許可。

以下狀況可不用申請：
1. 自有透天厝。
2. 樓層數 5 層以下的集合住宅不用申請，但若有「新增廁所及衛浴」、「新增 2 間以上的居室造成分間牆之變更」則需申請室內裝修許可。

Q10

買老屋好貸款嗎?

銀行近年對老屋貸款的審核,以及核准的貸款方案,雖然變寬鬆許多,但老屋貸款仍存在許多眉角,建議可多找幾家銀行做比較,再從中找一個適當的銀行進行貸款,以下是貸款前要注意的事項。

1. 老屋貸款條件
老屋貸款審核嚴格,且注重房屋本身條件,如:座落地點、鄰近設施、屋況狀態等。

2. 老屋貸款額度
一般來說,不是位在精華區的物件,老屋貸款成數約落在 5 〜 6 成左右。

3. 老屋貸款利率
考量到屋齡、折舊率及耐用年限等因素,老屋貸款利率和其他屋齡相比,會稍微高出一些,大約落在 3 〜 4%。

4. 老屋貸款年限
由於老屋剩餘年限不多,多會以屋齡和申請人年齡來綜合評估,並限制老屋貸款年限。

空間設計暨圖片提供│寬月設計

● 老屋因為屋齡與折舊關係,除非是在很好的市中心地段,否則貸款成數低,未來也不易脫手,想買老屋翻修,事前要想清楚。

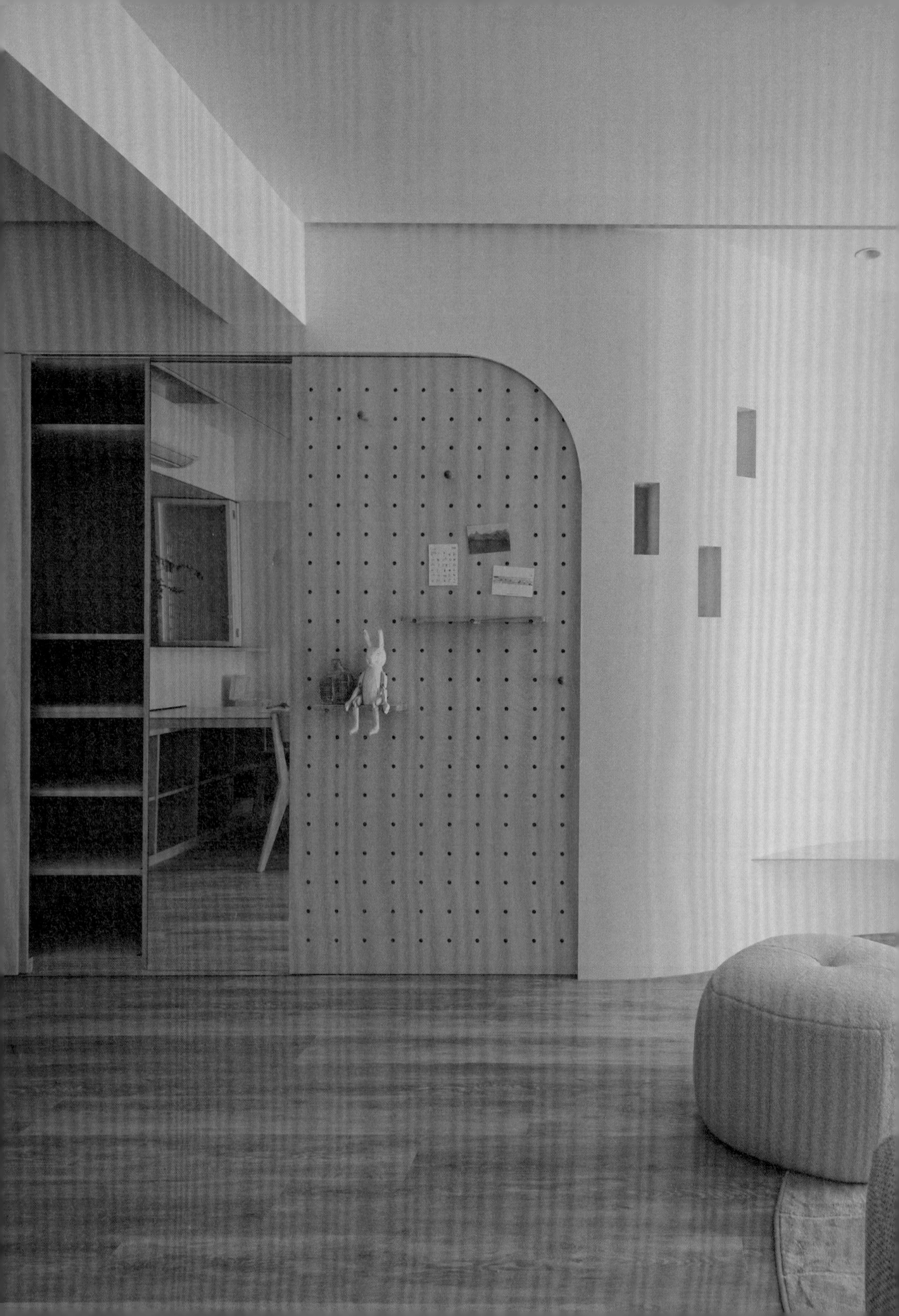

CHAPTER

2

老屋裝修,可以這樣做

Point 1 老屋工程這樣做

專業諮詢與圖片提供：日作空間設計、森叄設計

空間設計暨圖片提供｜日作空間設計

01 基礎工程

做好基礎工程，
再住 20 年沒問題！

　　對多數人來說，買房裝修是人生大事，特別是選擇買老屋翻新的屋主心中應該多少已有準備，畢竟老房子因時間與使用軌跡，或多或少會留下一些時光線條或傷痕，這些痕跡不只讓老屋更有底蘊，相伴而來的也是屋況在歲月洗禮後會形成舊損，而如何能讓老屋住得更舒適，「基礎工程」做得好不好是最重要的環節之一。這些工程包括有：拆除工程、水電工程、門窗工程以及地板工程，每一環節都是老屋裝修的不可或缺的重要基石。

●拆除工程應注意安全性，除了結構樑柱必須保留，剪力牆也不能隨便動，社區還需有大樓管委會同意才能動工。

購屋前應審慎評估時間與經費

對於裝修小白來說，喜孜孜地買下老屋，也準備拿出百萬來做裝潢，期待幾個月就能有完美的新家，但這樣的美夢往往在設計師這一關就被一盆冷水澆醒了。畢竟現在人工物料甚麼都漲，想做老屋翻新沒個幾百萬確實很難確實又完整地讓屋況修復。另一方面，老屋翻新不像新成屋或中古屋裝修只需 2～3 個月就可完成，從評估、討論到拆除、開工裝修，往往需要半年甚至 1 年以上的時間。因此，購屋前應該更審慎做功課，以免錯估形勢、陷入進退兩難之境。

老屋該不該作翻新呢？

老屋到底在甚麼狀況下需要翻新？專家說明大多數老屋都有複合性的屋況問題，常常是從局部來看可能還堪用，但有些地方又好像舊損不堪，就連有經驗的設計師都說很難給出明確答案，或許可以從多面向的評估來協助判斷。

老屋屋況評估表

1. 屋齡已經超過 25 年以上？
2. 老屋格局明顯與自己居住需求不合，例如房間數不足。
3. 廚房或衛浴設備老舊，使用需求無法滿足，或收納機能不足
4. 屋內牆面、天花板 (特別在浴室、廚房周邊) 出現壁癌、水珠或潮氣。
5. 牆面明顯可見有裂痕、縫隙或歪斜不正。
6. 地板磁磚出現有膨拱，或木地板有翹起或被蛀蝕現象。
7. 門窗框架或周邊泥作有明顯裂縫或水痕，或因框架歪斜而開關不順。
8. 木作已有被蛀蝕，或是門框壞損嚴重。
9. 過多隔間牆造成室內採光差、通風不良，以及感覺狹隘壓迫感。
10. 超過 70% 以上的舊式裝潢，或明確想換不同風格。
★若有以上問題請打 V，當 V 超過五個以上應考慮翻新，並諮詢專業團隊。

水路 & 電路工程

　　一旦決定為老屋重作裝修後，哪些是一定要做的工程呢？每天都會用到的水路與電路可說是最重要的，也與生活品質息息相關，更是未來二十年能否住得安全的關鍵工作，因此被列在老屋裝修基礎工程的第一順位。

老舊水路造成鏽、堵、漏，生活不便又影響健康

　　屋內水路工程主要鎖定廚房與浴室兩個水區，超過 20～30 年、甚至更久的老屋設備因為多半都已不合時宜，加上衛生考量，通常會採全部換新、重新規劃，尤其很多家庭的熱水需求量提高，可能會在廚房增加廚下型熱水器或瞬熱型飲水機，如果水壓不足恐怕無法安裝使用。因此，無論是加裝加壓馬達或是管徑需求都需要做變更，提升到 4 分或 6 分，甚至一吋水管，這樣才能因應新型用水設備的需求。

至於水路管線的重建則須區分為給水與排水兩個不同系統。

｡給水部分

早期的水管多使用塑膠管或鐵管，因耐候性較差，久了容易有老化破損、鏽砂堵塞，甚至長青苔的情形，破損的水管還會導致房子產生漏水、壁癌等問題，至於修繕的做法則可分為明管與暗管兩種。

1. 給水明管設計：較簡單的方式就是將原本埋在地板或牆內的舊管作廢封管，直接在牆面外拉新水管來使用，雖然較不漂亮，但方便施工，也深受許多工業風青睞。

2. 給水暗管設計：想以暗管方式修繕需要拆牆將原水管打掉重新埋管，這種做法可配合格局調整的地板與泥作工程一起進行，完工後較為美觀。

○ **排水部分**

老房子經過多年使用，特別是廚房排水管會卡很多油垢，導致管徑變小或堵塞，形成排水不順暢、回堵等問題，老屋修復團隊會先請專業的清潔工班來清理排水管，並檢查有無漏損狀況以保持排水管道暢通，也會叮囑屋主在日後使用要盡量排除油汙，讓裝修後的老屋可以住更久。有些老屋因排水共管有氣壓不平衡的問題，在一處使用馬桶時，另一處可能會有氣泡產生或水紋浮動，這些都須在裝修安裝時測試，並裝上通氣管來保持進氣平衡。

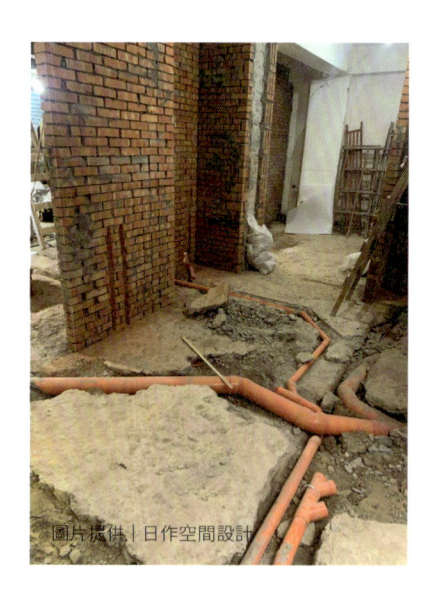

圖片提供｜日作空間設計

◉水路工程分為給水與排水兩部分，排水管須配合在新地板施作前先行配管施工，以免做好才發現有問題就須重做。

電路規格不符時代需求，新穎電器難使用

電路管材因為年久會老化、破損，使用了二十年以上的管路隨時都有可能會有電線走火、跳電等危險，所以會建議全面換新才能確保居住安全。特別是現代家庭電器暴增，與數十年前住宅的電力規劃不可同日而語，因此多數老宅重建時都須重新向台電公司申請增加總電量，電線及迴路配置也需要重新規劃。要提醒的是，一般公寓或社區會有共用管道間，若要增加用電安培數可能需要跟管委會詢問協調，如果是透天老厝則比較沒有問題，只需向台電申請即可。

總電量應提高：早期老宅申請的家用總電量可能只有 35 安培，目前一般家庭所需電量多在 50 安培以上，有些甚至會達到 70～75 安培，這些可依據自己的需求與設計師溝通來決定。

電路線徑換粗：重新規劃電路時，新建電路幹線的線徑也要更粗才能提高安全性；特別在廚房或者冷氣等重電家電則要規劃專用的獨立迴路，以免多種電器同時使用時造成跳電情況發生。

插座須充足：新電路規劃另一重點就是插座安排應充足，設計師提醒位置也要跟隨屋主的生活動線來作配置，這樣才能真正達到生活方便性。

地坪、牆面泥作

現代房屋建造過程有很多施工的輔助儀器，但早年師傅在造屋時多半是憑經驗傳承來土法煉鋼，容易產生一些地不平、牆歪斜的狀況；除此之外，有些老屋還有地勢傾斜或泥塊剝落、鋼筋外露，這些牆面重砌與地板整平重鋪的泥作工程嚴重影響美觀與安全，也是老屋翻新的重要基礎。

泥作修復是老屋延壽的關鍵工程

在老屋翻新時，泥作工程是地坪或牆面最大宗的施工項目，而且泥作工程常常無法一次進場就完成，而是必須配合其它如門窗、水電工程的進程來安排多次進場施工。泥作是房子的骨架，也是讓老屋能再多住個二十年，甚至更久的關鍵工程，所以務必要審慎評估再施工。

施工前可以先確認地面現況，通常老屋地板大多會刨除重做，但也有些老屋地板很有特色，例如早期磨石子地板，且屋主也希望保留，就可配合設計營造懷舊風格。不過，保留前還是應該先幫地板作健檢，以免翻新裝修後才發現地板內部的管線或材質有質變等問題，以下提出幾種老屋地板常見問題與修復方式。

空間設計暨圖片提供｜日作空間設計

●通常老屋地板大多會刨除重做，與此同時也能順便校正地面水平，但若屋主希望保留，原始地板狀況也堪用，便會保留下來，可配合設計營造成懷舊風格。

1. 如果地面只是凹凸不平或發生膨拱狀況，可在刨除後重整地面，並在新舖地坪或貼磁磚時運用整平器輔助施工卽可。

2. 老屋較令人擔心的是鋼筋嚴重外露的海砂屋，若遇到這種情況建議最好是呈報縣市政府來申請都更，而非強作老屋修復工程，以免後續更難處理。

3. 有些老屋因地震或老舊有牆壁或地面裂縫，若只是有明顯裂痕問題應該不大，但若是裂縫較大或鋼筋已有輕微外露，可以用樹脂填補修復，或先用防鏽漆處理再抹平重建。

4. 基地處於河川圳溝周邊的老屋，久而久之易造成地基沉降而產生高度落差，有些甚至高低可相差 10 公分以上，若確認無安全疑慮則可運用泥作重抓地面水平，後續牆面也才能順利施作。

5. 有屋主擔心老屋重建隔間牆會有承重問題，但這個問題主要出現在高樓建築，容易因隔間多、承載過重而出問題，至於老屋多半是低樓層公寓或透天厝，原則上只要維持原屋的隔間量就可以，所以採用砌磚牆也不會有問題。

6. 地板與牆面重建要考慮的問題還有水電管線與門窗位置，所以這部份的工序要特別注意。通常在泥作放樣後會先砌磚牆，接著請鐵工為窗戶、門檻作定位，之後再請水電配拉管線、預留插座，最後請泥作再次進場施工。

圖片提供│日作空間設計

● 老屋地板常有膨拱、破裂或傾斜等問題，為了打好基礎可在泥作地板修復時藉由新式儀器來重整拉平。

防水工程

老屋的防水工程可分為內外兩部分，室內主要是指廚房與衛浴兩大水區，戶外就是透天厝頂樓屋頂與公寓陽台、外牆，兩者的防水要點不盡相同。

○ 外牆防水工程

老屋的屋頂或牆面因年久導致防水層失效或裂縫產生，或是因水管破裂而有漏水狀況，這些問題若沒有積極處理的話最後恐變成壁癌，這樣一來室內裝修再好也沒辦法維持太久，所以外牆防水工程一定要做好。防水工程施作上應注意外牆必須先清潔、去除灰塵後再上底漆效果才會好，防水底漆要有彈性且能填補裂縫及強化轉角區，通常需上 3～6 道防水漆效果才更好。有些屋外的工程會有危險性，有些則需與其他住戶協商後才能處理，這部分設計師坦言不一定能盡善盡美，只能就自己的部分盡量做好外牆防水層。

廚房與衛浴區防水工程

室內防水工程較單純，過程也與一般新屋無異，在泥作完成待乾燥後塗上防水層，建議等 2 至 3 個工作天後，並確認漆乾了才可上第二層，重複三層後可以倒水進行測試，主要觀察防水牆的外圍泥作沒有潮濕或顏色變深就可鋪上磁磚或面材，同時注意浴室防水牆面高度需高於 180 公分以上才能達到防水效果。

圖片提供│日作空間設計

●防水工程主要在衛浴與廚房兩區，施作時在轉角接縫處須作強化，防水漆層則要施作三道，待全乾後再進行漏水測試。

門窗工程

如果是四十年以前的老房子多半有鐵窗，至於三十年前則常見黑色鋁門窗，這些老房子的窗戶或框架不僅讓房子在換新裝修時透露出老態；同時門框、窗框也會因地震、颱風，以及經年累月受風吹日曬、搖晃，氣密或水密性難免變差，也會有隔音性不佳的問題，為了維持一定生活品質，老宅在翻新修繕時多半都會選擇換新大門或鋁窗，這也成為老屋裝修的必要基礎工程之一。

窗框打除重作可一勞永逸

多數老屋門窗的翻新裝修都是配合格局變動與泥作一起施作，可以一併將窗戶滲水、漏風或者靜音度一次搞定，有效改善老屋居住品質。特別要提出來的是如果發現門窗的周邊牆面已有白華或是水漬痕跡，窗框牆面常見有泥沙、甚至泥塊脫落的狀況，即使短期間沒有見到漏水，還是會建議要將整個窗戶打除見底、重新換新框才能一勞永逸。

窗框施工時特別要注意，應仔細填縫來銜接原本牆體，如果可以在窗外作雨遮、防水漆等防護，未來防水效果也會更好。另外，若是牆體已嚴重到有鋼筋外露狀況則須以防鏽漆打底，再用砂漿、樹脂作填補，讓防水工程更完整，也避免新作的窗框不久後又鬆動、滲漏了。

●門窗修復或重建時，應注意窗框與牆體之間的密合度，特別是若有裂縫與剝落泥沙或碎石，應清除並補強。

圖片提供│日作空間設計

不動泥作可採乾式工法為鋁窗換顏色

但是也有些屋主買下老屋後覺得屋況還不算太差，格局也合用而不需重做隔間牆，然而年久的老式門窗的氣密性與水密性都已經變差該怎麼辦？一般可以敲掉窗框來更換新窗框，此時可同步檢查牆面交接處有無滲漏，並藉此機會作縫隙填補，然後依據自己對於戶外的嘈雜、風切或聲音的忍受度來選擇適合窗型。

如果不想破壞牆體、動到泥作，但又想換掉窗框顏色，以避免早期的黑框鋁窗讓畫面顯出陳舊感，鋁窗師傅建議可選用乾式工法，這樣就不必敲窗框，只需拆掉玻璃窗，再用新作的窗框來包覆舊窗框即可，可讓窗戶換上新色，但是缺點是窗框會變得寬一些，窗戶也會略小一點。

02 裝修工程

回春裝修，
讓老屋越住越回甘

老屋的基礎工程比較像是作完健檢之後對症下藥，把壞死組織、病灶剷除，再積極調整骨骼、補強營養，讓老屋得以延年益壽。但這無法讓老屋看起來更年輕體面，因此必須透過裝修工程，一如醫美手術來讓老屋回春。這也包括老屋與現代生活常有格格不入的問題，如格局繁複、樑位過低、面寬不足、採光窗型過小，或是動線過長或曲折等等，都需要透過現代生活的美學與時尚建材來作優化。

問題一、隔間牆太多導致空間陰暗

裝修解法1：
三、四十年前多代同堂很普遍、家庭人口也較多，因此老屋房間數通常不少，這就會造成隔間牆過多，且容易擋住採光。釜底抽薪的解決方法就是減少房間數，所幸多數新屋主都能接受刪掉不需要的房間，讓室內變得明亮、開闊些。

裝修解法2：
如果房間數無法減少，可以先檢視是否有房間可以採用開放設計，或是利用摺疊拉門等彈性隔間來因應，例如書房、和室或是客房等隱私需求度較低的空間，平日可打開讓光線進來，需要休息或隱私時再關上使用即可。

●老屋常因隔間牆多或光源單一，讓空間顯得陰暗，在改造時可以打開局部牆面，如作開放餐廚或重配格局來改善。

●無法減少房間數，可在房間與公共區之間立一扇玻璃窗，搭配拉簾使用讓採光可串聯，睡覺時也可關上保持隱私。

裝修解法 3：

變更牆面材質也是想要引入更多採光的常見手法，不少老屋改用玻璃磚牆取代實牆，透過磚材的漫射效果讓光線進入室內，又能體現老屋的溫柔氣氛。另外，老房子中常見的壓花玻璃也相當適合，常見的海棠花、長虹、格紋、水波紋等，都能幫裝修風格加分。

問題二、樑位低、巨柱矗立引發壓迫感

空間設計暨圖片提供｜日作空間設計

●開放式格局設計讓客餐廚共享雙面採光，過低的天花板則改以筒燈與軌道燈來設計，拉升屋高來減少壓迫感。

裝修解法 1：

早期房子屋樑較低，還有前面提過老屋改造時常會將隔間牆拆除，但礙於結構安全考量，牆面拆除後仍必須保留大樑或柱體，這些低樑巨柱就容易讓空間有壓迫感。最簡單的解方就是透過貼鏡方式來包覆樑柱，讓樑柱反而成為反射空間的利器，延伸視覺與場景，營造更高挑或更寬的空間感。

裝修解法 2：

樑柱配合風格作成特殊造型也是很不錯的化解方法。例如利用樑柱來設計最近很流行的拱門造型，不僅設計合理，也能展現出風格特色。如果不想過多造型，也可運用導圓或斜切設計簡單帶過，或是以階梯式設計來減緩天花與大樑的落差感。另外，如果是歐式鄉村風可用木皮包覆大樑，並且在周邊裝飾成排的枕木來強化風格元素；若是古典風則可作成宮格設計來掩飾大樑。

裝修解法 3：

若是柱體是靠著牆面的話，可以考慮利用假櫃設計來包覆柱體，讓整體牆面看起來簡約、平整。但是如果柱體不靠牆，而是矗立在空間中的情況則可在格局規劃時看看是否能將柱子視作區域的定位點，常見的就是餐廚區吧檯靠著柱子來設計，讓原本無靠山的中島也更有安定感。

●拆除牆面可能讓樑柱直接裸露於空間中，不妨利用樑柱做為空間的定位點，在開放格局中使書桌與沙發更具安定感。

空間設計暨圖片提供｜日作空間設計

狹長格局、過長走道造成空間浪費

裝修解法：

傳統街屋或老公寓很常見到前後採光的長屋格局，除了採光受阻，因為隔間多更會形成長走道格局，除了減少房間數外，有經驗的設計師也會跳脫傳統想法，給予不同的格局建議，例如將原本客廳與餐廳位置對調，看能不能有助於打開新的局面。或者是在長格局的中間位置安排天井或是開放的和室之類，這些顛覆屋主想法的格局設計有機會讓過長走道被化解，以減少空間浪費、提升利用率。

前後陽台、樓梯下畸零空間規畫

裝飾解法1：

不少老屋的前後陽台因為髒亂老舊而被廢棄，在高房價年代更顯浪費，所以趁翻新時可將前陽台好好利用，如果在一樓可以規劃為綠植庭院，若是在樓上的前陽台也可在法規許可下拆除鐵窗重新規劃，一來可提亮室內光線、增加收納設計，或改造成玄關，打造雅致的落塵區也不錯，至於後陽台則可整理為舒適的曬衣工作區或儲藏區。

裝飾解法2：

複層建築需有樓梯才能連結上下樓層，因此會產生梯下的畸零空間，除了可作為收納室使用，樓梯本身也是老屋改建的重點，以往多半是厚重的泥作樓梯，如果改成螺旋梯除了省空間，優美造型也能成為屋內焦點。或是改以輕盈的龍骨梯或懸臂梯來減少量體、避免遮蔽光源，搭配鐵件材質更輕盈、也符合現代空間風格。

空間設計暨圖片提供｜日作空間設計

●傳統樓梯多採用泥作，顯得笨重又遮蔽採光，改建時可改以鐵件龍骨梯搭配玻璃護欄，讓空間更明亮輕盈。

留用老屋元素，點燃創新火花

裝修解法：

為了有更好居住環境，老屋翻修時常常是全屋拆除重建，但如果老屋的硬體結構狀態還不錯，且具有文化美感也可作保留。特別是遇到喜歡懷舊元素的屋主，可考慮留用局部結構或建材，反而能凸顯特色。常見留用元素如磨石子地板、老磚牆、壓花玻璃窗、鐵窗花以及鐵件樓梯扶手，這些物件除了可以原件保留，如果略顯舊損也可稍作修復或者調整色調後，再視情況融入新家。另外，如果屋主想將舊家家具留用到新家中，建議可由舊家具的色彩與風格來延伸調整出新的空間配色，降低舊家具的違和感，讓老屋空間也能住出回甘新滋味。

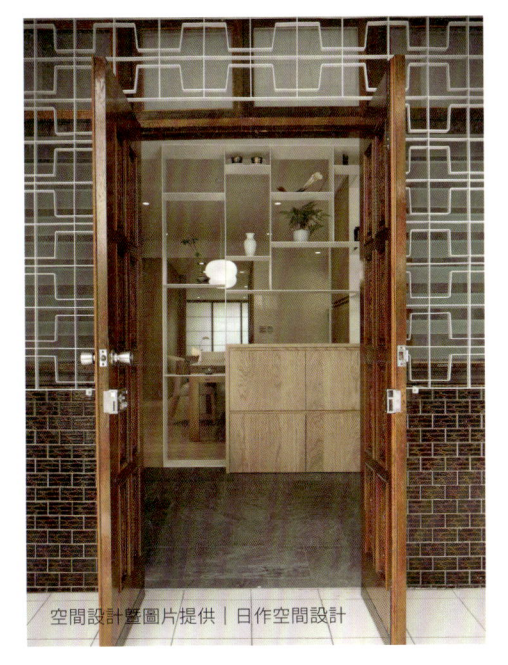

空間設計暨圖片提供｜日作空間設計

●由於老屋仍保有典美的木窗框、磨石子樓梯，所以選擇類似色調的灰磚與白色塗料牆來呼應，彰顯老屋懷舊氛圍。

Q & A

Q1

明明全部打掉重做，
但還是出現漏水問題，是工程沒做好嗎？

很多人以為，老屋翻修只要一切按部就班照著做，接下來就沒問題了，其實要有一個認知就是建築本身結構會自然老化，老屋翻新只是將因為時間而變舊、毀壞的地方做修繕與換新，但有些問題並非可以完全根絕，像是漏水問題，除了可能是自家問題，也可能來自樓上或隔壁鄰居，又或者當時進行工程時沒問題，是裝修完成後才發生漏水。

比起新成屋，老屋不可預期的狀況本來就比較多，也因此老屋不只翻修過程難度高，也要承擔可能突發不可預期狀況，因此不是翻修完了就一定沒問題，要有可能會出現需要事後維修的心理準備。建議可先請原設計公司到場堪查，並確定漏水問題，究竟是工程問題，還是來自周遭鄰居，找出問題根源，再進行後續維修。

Q2

裝修老屋時，哪些舊有元素可以保留使用？

老屋翻修也不是所有東西都不能用，全部都要拆除換新，如果地板損毀情形不嚴重，早期的地板磁磚、磨石子地其實很有具時代特色，若能保留下來，也可當成空間裡的一大亮點，畢竟地板若要全部重鋪，也是一筆不小的費用。另外，原始建築的磚牆若狀況良好也可以保留，因為極具古樸手感的磚牆，不論是在古典還是工業風空間，都相當適合。至於在老建築才看得到的老窗花，也是很適合保留下來的設計，加以規劃融入新建築，可為建築外觀增添更多亮點與特色。

Q3

房子二十幾年了，看起來問題不大，一定要花大錢翻修嗎？

二十幾年的房子，算是資歷最淺的老屋，正常情況下，這個年份的房子建築結構應該還很堅固，外觀保養得好，問題應該不大，是屋況算好的老屋，以這樣的條件想省錢簡單翻修，主要還是要看幾個可能花大錢的工程是否會因屋況費用再追加。首先不能省的是水管、電線的更新，過去用電方式和現代人用電習慣大不同，只要超過二十年以上的房子，建議最好全室水管、電線重拉，而且用了十幾年，可能都已有損毀、鏽蝕狀況，若不更新，恐會危及居家安全。

另外是一動可能要花不少錢的地板，如果損毀嚴重就要拆底重做，但若耗損程度低，與規劃的室內風格也相符，可簡單修整即可。再來是門窗結構是否完好，如果材質堅固耐用，表面漆膜完好可保留，二次施工時，貼上裝飾面板符合裝修風格即可。最後是要動到拆牆的格局規劃，基本上只要不大動格局，費用就會明顯少很多。

其實老屋翻修一般工班並不太願意接，不只是因為費用問題，而是可能會涉及工班無法處理的結構問題，所以不建議自己發包裝修，最好找對老屋翻修有經驗的設計師會比較有保障。

空間設計暨圖片提供｜十一日晴空間設計

● 不論是否想大幅翻修，老屋翻修還是有一定要做的工程項目，像是水管更新，其它工程則視實際屋況，及個人需求，再決定是否花大錢翻修。

Q4 所謂的基礎工程是指哪些？

基礎工程佔據老屋裝潢費用最大宗，但基礎工程到底做了些什麼，又包含了哪些工程？以下列出基礎工程包含的項目：

。拆除工程：主要是在拆除房子舊有裝潢，與新成屋不同，老屋的拆除工程較為繁雜，而且要小心有些承重牆和樑柱不能拆。

。水電工程：針對老舊、破損的水管、電線做更新，尤其是超過 20 年的老屋，多半都應該全室換新才安全。

。泥作工程：主要是砌牆面，或者是地板打底重做，也會需要進行泥作工程。

。防水工程：針對防水、漏水問題，進行抓漏加強防水，確保全室不會有漏水情形。

。油漆工程：在漏水、壁癌等牆面問題解決後，便會批土讓牆面變得平整，接著再上底漆和面漆。

空間設計暨圖片提供｜賀澤設計

● 老屋想要再住更久，最重要的就是把基礎工程做好了，之後才能住得舒服又安心。

Q5

牆面出現壁癌，可以直接批土油漆處理掉嗎？

老屋牆面翻新最常見的就是重新刷漆，一般人認為只要直接將新油漆塗上即可，但事實上老屋的牆面通常會有舊漆剝脫、牆面不平整問題，若是嚴重一些，牆面可能都已出現滲水，甚至成為壁癌，那麼只是在表面塗刷油漆理掉並無法根治。

若只是牆面油漆剝落、不平整，那麼做法是磨掉舊漆，重新批土抹平牆面再上新漆，針對有滲水與壁癌的牆面，則是要將牆面打除至見底，然後再重新做防水及打底粉刷油漆。

Q6

舊家具應該要保留嗎？

在進行老屋翻新時，屋內的裝潢大多會全部拆除，而此時舊有家具去留就成了一大難題，因為有些家具雖已不符合時下審美，但因為品質不錯，想丟棄又有點可惜，此時要怎麼解決比較適合？

首先，可透過微改造來讓舊物展現新風貌，例如重新油漆，或者貼覆新的表面材質，將原本不符合流行的樣貌做改變，再來也可以根據風格，更換沙發套，或添加一些織品，來讓家具變成另一種樣貌。

另外，若家具狀況不錯，也可以選擇在 FB 社團、網路買賣平台二手拍賣，或是捐贈給救助弱勢的單位或平台，例如：iGoods 愛物資、心路基金會、慈懷社福基金會、伊甸社福基金會等，捐贈前可以先詢問是否有缺該項物品。

Q_7

因為拆牆留下一根柱子，怎麼設計才能看起來不奇怪？

在老屋翻修過程中，經常會因為要重新規劃格局，而進行拆除隔牆、樑柱，空間雖然因此變開闊了，但如果遇到不可拆除的樑柱，就會形成空間裡相當突兀的存在，而且也容易產生不好利用的畸零地。

首先，想化解這根柱子，可先根據空間風格，在樑柱上貼覆鏡面、木素材等表面素材，藉此來美化樑柱外觀，弱化柱體存在感，而且鏡面材質，還有延伸視覺效果。

再來可將樑柱結合收納設計，把柱體收進櫃子裡，這樣一來視覺上就簡約許多，而且也因此多了收納空間。

若是因為樑柱關係，而產生了難用的畸零地，在進行收納設計時，可一併規劃進去，若空間比較大，則可利用臥榻、桌板、平台等設計，來解決形狀不規則的畸零地問題。

若是樑柱是位在天花板，此時建議可直接裸露，搭配明管設計，不只能提高樓高，同時也能呈現較為粗獷、隨興的工業感。

空間設計暨圖片提供｜福研設計

◉老屋除了常見樑柱太多，還會因為在翻修過程中，拆除牆面而裸露出不可拆的樑柱，此時可透過設計，讓樑柱的缺點轉化成優點。

Q8

正要進行老屋翻修，地板看起來好像狀況不錯，一定要拆掉重做嗎？還是有什麼其它方式可以處理？

老屋雖然老舊，但也不是所有裝潢都要拆除不能使用，像是地板就能根據老屋的實際地板狀況，採取不同程度的翻修，藉此也可能可以省下一點費用。

◦ 保持原狀
若老屋原屋主有定期幫地板或地磚做保養，那麼耗損程度應該會比較低，此時先確認地面破損狀況是否嚴重，若不嚴重，且也能與空間風格符合，那麼可以不用全室重做，只要簡單修整就可以了。

◦ 局部翻新
若只有局部破損，那麼翻新時針對該區域做修整，不需要地面全部換新，例如：地磚少數缺損，購買相同磚材補上，若為木材質，先確定是實木還是超耐磨木地板，若為超耐磨木地板，無法修整只能換新，若是實木地板，便可使用機器重新打磨表面，再漆底漆、表漆、打蠟，讓地面恢復樣貌。

◦ 全部翻新
如果地面已有嚴重破損，且表面磨損嚴重，地面應該全部翻新。一般會拆除見底後，再根據地板使用的是磚材、木地板等建材，來進行後續地板舖設的工程。

空間設計暨圖片提供｜日居設計

●老屋地板大多有不平的問題，因此地板的更新，除了是修復破損目的，還要將全室地坪調整成一致水平，因此不能只從美觀考量是否換新。

Q9 老屋哪些一定要做？哪些可以不做？

想更精準的掌控老屋翻修預算，就要知道在翻修過程中，哪些是一定要做，哪些又可以視狀況不做，以下列舉出一些項目。

1. 防水
老屋最大的問題就是漏水問題，因此防水工程不能省，而最常用水的兩個區域就是衛浴和廚房，因此這兩個地方不論現況如何，都建議拆除重新翻新，在防水處理上，要注意牆面轉折處、材料銜接介面。

2. 水電重拉
線路也會老化，因此二十年以上老屋最好重拉。

3. 更換窗戶
窗戶不夠密合也是容易導致房屋滲水的原因之一，因此建議要更換，只是可視原始狀況與預算，選用不同工法來施工，且可考慮更換具有防風、隔音、隔熱功能的氣密窗。

4. 壁癌漏水處理
若有壁癌問題，一定要從根源做處理，否則後續牆面將會難以進行上漆、貼壁紙等工程。

5. 地坪更新
地坪可視原始狀況，選擇是否更新，若沒有不平問題，破毀問題不嚴重可保留，或局部更換，省下一筆費用。

6. 舊家具
老屋翻新時，原來的舊家具可以再利用，只是可能需重新整理一番，以符合現代美感與空間風格。

7. 隔牆拆除
若老屋格局沒有太大問題，其實也可以考慮不動隔局，如此一來隔牆也不用拆除。

Q10

老屋翻新流程和新成屋一樣嗎？
主要有哪些流程？

老屋翻修流程和新成屋差異不大，但施作流程會隨著屋況、預期翻新的程度而有所差異，會進行的基本流程如下：

1. 保護措施與防護工程
於公寓、大廈的翻修工程要做好保護措施，除了在走道、電梯鋪設防護軟墊，以免工具、建材進出損傷公共設施外，可能需與大樓管理處告知，於公佈欄張貼翻修公告，知會左鄰右舍。

2. 泥作與水電配置等基礎工程
從拆除、配管、天花板、地板，到貼壁磚鋪地磚等，都屬於基礎工程，也是老屋翻修主要工程。此階段相當於為房子重打基礎，因此是老屋翻新費用中較高的預算項目。

3. 裝潢工程
工程內容包含木作施工、天花板、隔間牆面、油漆等，此階段多是依照個人喜好和需求，是可以趁機節省一點費用的工程階段。

4. 軟裝
家具、系統家具、壁紙、窗簾等，主要在美化空間，讓居住者可以住得更舒適。

空間設計暨圖片提供｜佳設計

●新屋和老屋裝修流程差異不大，只是會因屋況與翻新程度，而讓工程比例與工期長短略有不同。

Point 2 破解老屋格局問題

空間設計暨圖片提供│賀澤設計

01 缺少採光

開放與通透隔間，
搭配淺色、反射材質提升光線

　　台灣許多老屋由於早期建築多為連棟設計且偏向長型、封閉，導致容易造成室內光線不足，影響舒適度。然而透過格局調整、透光隔間或拉門的運用，以及增加淺色或鏡面材質，皆可有效改善光線不足的問題，變得更加明亮舒適。

調整格局讓光流動也製造開闊感

　　許多老屋格局都過於封閉，空間被多道隔間牆分割的情況下，使得光線被阻擋在某些區域，自然導致昏暗甚至無光。若是因格局規劃不當造成缺少採光，建議可以適當將格局重整，拆除非必要的隔間，使光線能夠自由地流動，同時建議將主要活動空間安排在採光較佳的區域，像是藉由開放式設計將客廳、餐廳與廚房、書房整合在一起，如此不但可以改善光線流動，也能讓視覺變得更加寬敞。

　　除此之外，若場域之間必須得有區隔性，也可以改為採用半開放性的規劃方式，例如半高牆、開放式櫃體或是活動隔屏拉門等形式，同樣可維持光線的通透與延伸。另外針對老屋走道過於狹長與陰暗，則可透過局部鏤空或是採用玻璃拉門的方式，讓光線能夠順利進入走道，減少暗區的產生。

運用透明與半透明材質，讓光線自由穿透

　　面對無法捨棄的隔間，可以改用透光玻璃材質來提升室內光線的流通性。這樣的做法多半出現在像是書房或是多功能房，可利用如玻璃隔間、拉折門取代實心牆，既可讓光線延伸到其他空間，同時也可以保有獨立性，其他像

是噴砂玻璃、壓克力和透光磚等半透光材質則是能在維持隱私下，同樣讓光線自然擴散。其次，開放式層架也是提升採光的好選擇，取代封閉式收納櫃後，可減少光線遮擋，使整體空間更加通透明亮。以上這些設計手法不僅能夠改善採光，也讓老屋的空間感更顯輕盈寬闊。

善用材質與顏色，提升光線反射效果

除了格局調整之外，適當運用材質與顏色，以及掌握軟裝細節也能有效提升光線，使空間更顯明亮寬敞。從材質面來看，像是鏡面或亮面磁磚、烤漆處理的櫃體，都能夠透過光線折射，創造透亮的視覺效果。舉例老公寓玄關處通常較難獲得採光，此時可選擇於牆面安裝灰鏡、茶鏡等，透過反射進一步提升亮度、放大空間尺度，更兼具外出整裝的作用。

空間色彩選擇方面，牆面與天花板可多採用白色、米色、淺灰等高反射率色系，使光線均勻擴散，減少暗區的產生；地板則建議使用淺色木地板、拋光磁磚或亮面石材，以增強光線的折射與流動。軟裝部分，可選擇透光性較高的窗簾，如輕透紗簾或亞麻布料，能減少遮擋並保持光線的通透性。即使是家具的搭配運用，也可利用玻璃桌面或帶有光澤感的設計，使光線在室內形成更均勻的擴散效果。透過這些材質與顏色的整體搭配，即使原本採光條件有限，也能有效提升光線，營造出更加明亮、舒適的居住環境。

●面對光線較弱的廊道，電視牆面特意不做滿，包括層架也採用鏤空方式，讓光線能延伸擴散至室內。

空間設計暨圖片提供｜日居設計

空間設計暨圖片提供｜日居設計

◉長型老公寓常見的問題就是走道陰暗無光，為此設計師特別將廚房隔間改為玻璃拉門，可彈性開放、獨立的設計，平常維持開啟既可放大空間，也有助於提升明亮與舒適性。

空間設計暨圖片提供｜日居設計

◉除了改變格局，還可以選擇使用透光性強的玻璃取代實牆，搭配淺色木地板、白色牆面天花的顏色運用，同樣能達到提升光線反射的效果。

02 動線不順

避免狹長陰暗，
讓家的動線明亮又順暢

很多老屋的動線，就現在來看會覺得難以理解，這是因為過去的家庭結構與生活習慣，與現在有著很大的差異，過去家中人口多，希望隔出更多房，導致生活空間被切割得相當零碎，不只每房空間不大，還因此產生狹長、崎嶇的廊道，而這些不合理的廊道，現今只要透過適當的規劃，便能重塑空間動線，符合現在的生活型態。

減少隔間，避免切割空間形成長廊

過去家庭人口較多，且大多是多代同堂，現在則以小家庭為主，家中人口變少了，房間需求量變少，隔牆自然可以順勢減量，至於一些比較不需隱私性的公共空間，如：廚房、書房等，建議可以採開放式格局規劃，沒有實牆隔間，便能最大程度釋放空間，製造寬敞開闊感，沒有過多隔間也能減少廊道的形成，且有利於改善因空間被隔牆壓縮，隨之而來的壓迫、封閉感，也有助於解決老屋缺乏採光問題。

若是真的有隔間需求，像是書房、工作室需要獨立安靜，或擔心廚房油煙溢出，可以採用可彈性收放的拉門、折門設計，方便因應不同情境使用需求，同時又能滿足隔間要求，材質可依空間風格與需求選用材質，其中玻璃材質因具通透特性，不只有助於化解隔牆狹隘封閉感，還能達到引入光線目的。

空間設計暨圖片提供｜福研設計

● 由於原始空間較小，又有樓梯阻隔在中央，影響動線，空間也被切割零碎。因此樓梯往客廳挪移，有效分隔通往餐廚與上層臥室的動線，不互相干擾。同時樓梯多了轉折設計，能結合電視牆與收納，賦予靈活多元的機能。

動線取直，空氣流通住起來更舒適

　　老屋最常因為不當的隔間，而造成行走的廊道變得彎曲崎嶇，在進行翻修時會建議盡量將這樣的走道取直，因為彎曲的走道容易形成空間浪費，同時也不利於空氣流通，而空氣不流通、滯留情況如果嚴重，則很可能會衍生發霉等問題，若實在無法取直，像是雙動線、弧型動線都是能避免產生彎角的設計。

走道尺寸做對了，動線自然就順暢

　　除了常見的狹長、陰暗問題，老屋走道最大的問題，就是有的太窄有的太寬，如果狹窄到只容一人通過，則行走動線會不順暢，太寬又過於浪費空間，然而走道的功能就是用來串聯各場域，想讓家的行走動線更順暢，就應該要有合理且舒適的走道尺度。

一般來說，動線的規劃設計大致可分成主動線和副動線，主動線是較多人會行走的動線，像是從大門走到客廳、餐廳等，副動線則是較少人會行走的動線，如廚房、書房或臥室。大門玄關要預留開門空間，最好不要小於100公分，這樣就算多人同時進入空間也不擁擠，從客廳轉入私領域的走道，約90～110公分左右，讓兩人錯身時還有空間裕餘，副動線使用頻率較低，大致上只要不低於60公分，使用上都不會顯得侷促，不過若是兩側規劃有櫥櫃，則走道寬度要再加寬。

　　另外，兩人居住的臥房，盡量在床的兩側留走道，避免貼牆擺放，以免其中一人下床時影響到另外一人。廚房和餐廳的走道規劃，則與櫥櫃息息相關，最好根據櫥櫃深度進一步調整走道加寬幅度，一般廚房走道約90公分即可。

空間設計暨圖片提供｜樂治設計

●走道盡量維持筆直，不要有太多彎曲，且盡量不要形成狹長走道，若無法避免則要納入採光條件，以免走道過於陰暗。

03 空間不方正

順應地形，
以設計巧思化缺點為優點

　　早期台灣常見有外觀細長、扇型或圓弧型的房子，外觀雖然看起來很有造型又美觀，但由於建築本身是特殊形狀或不規則型，在進行空間規劃時，不只難度高還容易浪費空間，雖說可透過設計來解決，但也可能讓裝潢費用增加，因此建議盡量避開這類型老屋。

畸零地也能變身好用空間

　　不規則的空間，除了難規劃，最大的問題就是容易浪費空間或產生畸零地，想活用每寸空間，同時又拉直空間線條，最常見的做法，就是規劃收納櫃，藉由將不規則的線條收在櫃體中，便可達到校正空間目的，同時營造俐落空間感，與此同時還能增加收納空間，不過缺點是會因為收納櫃體，而犧牲部份使用空間。

　　全部都做成高櫃，可能會讓人到很壓迫，那麼可以根據空間條件，採用高櫃和矮櫃交錯搭配的設計，或是以矮櫃取代高櫃，選用的材質、顏色，建議最好挑選白色、淺色系等，來達到輕盈、弱化量體存在感。至於喜歡展示收藏品的人，可以選擇開放式層板設計，單純架設層板，費用也會比打造一座櫃體來得便宜。

　　有時因為格局、樑下或無法避開的樑柱，而形成不好運用的畸零地，此時有幾種做法，一種是在樑柱之間架設層板，不只可用來收納、擺放物品，同時也能結合樑柱形成一個立面，化解空間裡出現樑柱的突兀感；若很在意

樑柱，則可選擇規劃櫃體，把樑柱收進櫃子，視覺上相對俐落許多。遇到不規則形狀的牆面，可沿著牆面形狀打造懸空桌面，依空間屬性可當成梳妝桌、書桌。除此之外，也能規劃成臥榻，臥榻架高高度可隨個人喜好與需求，若有收納需求，還可在臥榻下面規劃成收納空間。

利用格局規劃與採光，弱化歪斜不規則線條

　　利用收納櫃雖然能讓空間變得方正，卻也會因此犧牲不少空間，而且打造櫃體的費用也不少，若不想採用這種方式，那麼就要依賴設計師以格局規劃來化解。若比較沒那麼在意歪斜、不規則線條，可以將這些牆面劃到使用空間最大的公領域，在寬敞的空間裡，不方正的線條不易被強調，也能藉由容易適應屋型的活動家具搭配，來轉移空間視覺焦點。至於在意的人，則可以將主要空間導正，剩餘的畸零地則做為使用頻率較低、時間也比較短的空間使用，如：衛浴、書房、儲藏室等。

空間設計暨圖片提供│季沃設計

●巧妙將兩個大樑柱結合多種收納形式，來形成一個完整的收納空間，且透過層板規劃，除了多了收納功能，還能保留空間裡的採光。

實例示範

拆除封閉隔牆，釋放採光與空間感

原始廚房封閉侷促，也阻擋採光。因此廚房拆除隔間，改為開放設計，與客廳、餐廳相連，釋放整排大窗的光線，還原空間深度。整體空間不僅流動輕盈，視野也更明亮開闊。空間設計暨圖片提供｜福研設計

半牆、橫推拉門引光，提升餐區明亮度

由於主要採光較弱，經過調整隔間配置，退縮第一間房深度後，同時讓遊戲室的隔間為半牆搭配玻璃材質，加上橫推拉門規劃，讓房間光線可流動至公領域，提升用餐區域明亮舒適性，同時也維持空間的開放與延伸效果。空間設計暨圖片提供｜十一日晴設計

開設採光窗，提升室內明亮感

老公寓中段因缺乏自然採光，將原本的房間改為主浴，並在臨陽台的隔間上方增設採光窗，引入外部光線，使空間更加明亮。同時，鄰近洗衣間的隔間在高度兩米以上也設置採光窗，進一步提升光線穿透效果。空間設計暨圖片提供｜十一日晴設計

開放式廚房，提升空間連結性

原本廚房位於屋內深處，距離客廳遙遠，動線迂迴且為封閉式設計，影響使用便利性。將廚房向客餐廳區域移動，並拆除原有隔間，改為開放式設計，讓三個區域得以串聯，提升空間互動性也讓公領域動線更為流暢。空間設計暨圖片提供｜十一日晴設計

畸零空間變高效家事間與儲藏區

老公寓原本三角形的畸零空間因不方正且動線不佳，過去作為衛浴使用較為不便。翻新後改為實用的家事空間，規劃為洗衣房並整合換季衣物收納與折衣平台，不僅提升使用效率，還兼具儲藏室功能。空間設計暨圖片提供｜十一日晴設計

注入設計巧思，畸零地也能發揮功能

主臥因無法消除的樑柱，產生難以利用的長形畸零地，不以實牆隔間，而是以大形衣櫥、矮櫃與鏤空設計，來形成立面，一面做為床頭背牆，另一面則用來收納衣物，再利用回字動線串聯，來確保動線流暢與採光。空間設計暨圖片提供│季沃設計

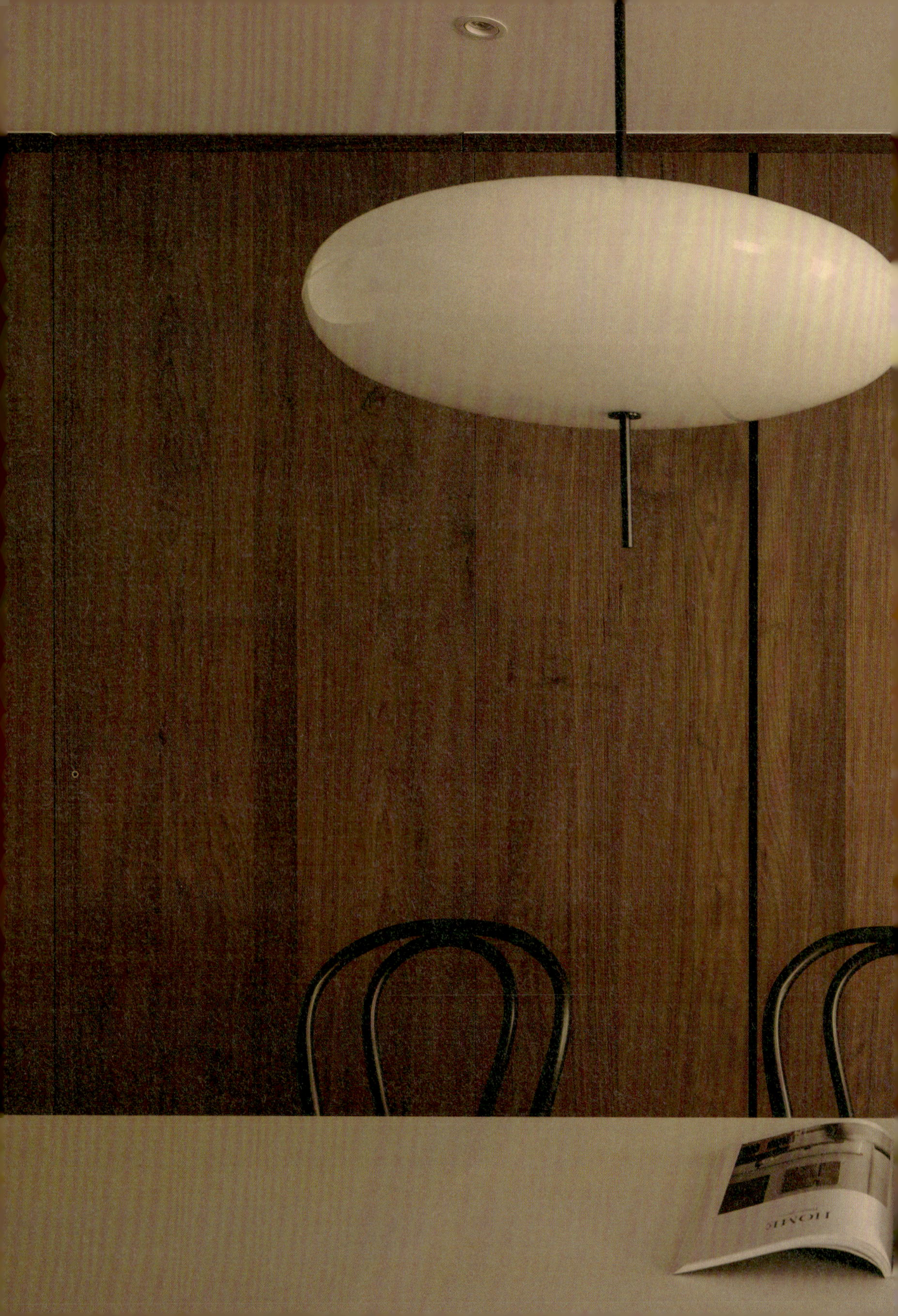

CHAPTER

3

空間實例

CASE 1

HOME DATA

屋齡｜30 年　屋型｜電梯大樓　坪數｜23 坪（不含陽台）　家庭成員｜夫妻、1 貓
建材｜水磨石磁磚、木紋磚、海島型木地板、實木集成材檯面、實木桌板、無毒礦物塗料

拆牆重塑動線讓生活變流暢

空間設計暨圖片提供｜日作空間設計　文｜Fran Cheng

●基地僅有前後採光，兩側幾乎無開窗，所以將後方廚房三個窗口搭開引入光源，並選擇打開廚房牆面改以吧檯作中介，藉由開放格局以及白色空間來襯托木質調家具，鋪陳日雜風居家感。

Before　　　　　　　　　　After

　　屋主是養著一隻貓咪的夫妻，太太是介面與平面設計師，先生則是喜歡古家具與老件的布料經營者，倆人因喜歡日雜風的木質感空間，加上對空間有滿多想法，所以陸續已買了些家具與燈飾，特別是他們也不急著購齊，而是隨生活步調再慢慢添購。

　　這是近三十年中古屋，翻修工作除了重整屋況，同時考量前陽台雖面向社區中庭花園，但後方三個窗口視野卻更開闊，為了讓後段採光可長驅直入，拆掉原廚房與次臥隔間牆，好放寬廚房格局且改採吧檯做開放設計，再搭配餐區木圓桌、玻璃吊燈，一旁加入屋主喜歡的拱門元素更顯懷舊氛圍。

　　餐廳一旁配置有咖啡 Bar，讓在家上班的屋主可在餐桌開會、辦公，呈現舒適的居家工作環境。

　　屋主認為臥室的重點在睡眠，無需衛浴間，因而將主臥移到右後的房間，至於主衛浴間則與更衣室串聯獨立在另一側，讓洗浴梳妝動線更流暢，臥房隱私性也更好。

三道間接照明營造柔光客廳

為改善客廳亮度，除了採開放式規劃來納入前後窗採光，考量結構樑柱大且低，所以天花板不挖燈孔，而是以三道間接照明來營造柔光環境，搭配透光不透景的白紗簾，打造有隱私卻明亮的起居空間。

玄關拱門鏡反射明亮角落

將門口小孩房與衛浴間門框融入屋主喜歡的拱門設計元素，除了為空間帶來軟化視覺的圓潤效果，小孩房拱門外側貼附鏡面可當穿衣鏡、延展出寬闊視野，映射落地窗光，讓玄關拱門鏡角落更明亮聚焦。

吧檯與餐區也滿足工作需求

將廚房格局放寬,且以木吧檯取代牆門作中介與界定,讓夫妻倆人可面對面互動外,冰箱旁還增設咖啡Bar,屋主可以在此沖泡咖啡,親挑的木圓餐桌與復古吊燈則可提供用餐、喝咖啡與工作使用。

梳妝、沐浴、更衣、清潔一條龍

屋主接受新觀念,將家事工作間、洗衣區、衣帽間、洗手台、馬桶間、淋浴間都串連在同一條動線上,藉由一條龍通道促成順暢的生活動線。加上動線上有多道門可控制通透度、增加使用的彈性度。

CASE 2

HOME DATA

屋齡｜約43年　屋型｜公寓　坪數｜23坪　家庭成員｜夫妻
建材｜油漆、超耐磨木地板、鐵件、烤漆、系統板材

釋放封閉廚房，讓長形老公寓滿溢日光

空間設計暨圖片提供｜樂湉設計　文｜陳佳歆

● 在公領域展開原來的封閉廚房後，將客廳、餐廳及廚房串連在一起，以簡約線條搭配暖白色，再以溫潤木材質點綴，傳遞出簡約卻不失溫馨的空間氛圍。

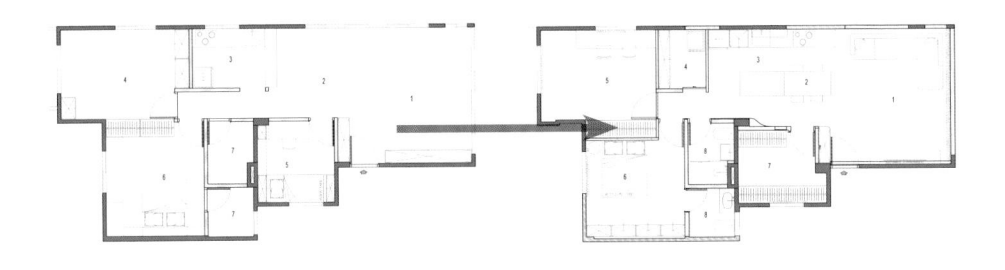

Before　　　　　　　　　　　　　After

　　年輕的夫妻北上定居，為了在預算與坪數之間取得平衡，選擇了位在二樓的四十年老公寓。

　　早期的狹長形空間，不但屋況老舊有些壁癌，樓高較低外還有不少橫樑，且封閉式獨立廚房空間明顯不足，這些都是待解決的問題，空間設計除了改變現況之外，也希望創造出居住質感。

　　雖然屋況老舊，所幸前段區域有L型轉角開窗，設計上延伸這項優勢，將窗戶改為大面窗減少線條，帶入充足光線；格局上釋放原有封閉的廚房，打造一個結合餐廳的開放空間，整個公領域都能感染明亮的日光，餐廳區域以弧形牆面柔化空間，同時形成一條引導動線，長窄的廊道在視覺上也更顯寬闊。

　　天花板為了儘可能保留原有樓高，採用局部造型修飾樑體，餐廳區以圓弧造形包樑不但呼應牆面流動線條，同時也界定了區域。

　　廊道底端的門扇特別選了小冰柱玻璃，讓書房的光線能夠從長型空間的另一側微微透入，中段空間因此不至於太過昏暗。

　　現代簡約的暖白空間中鋪上溫暖的木地板，圓潤燈具及家飾增添家的放鬆氣息，在忙碌的工作之餘，讓家的溫度成為夫妻倆最美好的生活回憶。

曲面造形牆擴大餐廳和廊道

廚房區域藉由曲面造形柔化空間，同時擴大過道及餐廳的尺度，更衣室採用內縮門片加上壁龕設計，為簡約的白牆堆疊豐富的視覺感受。

調整轉角大開窗視感更乾淨

原有的傳統舊窗戶寬幅較小，看起來有許多分割線，重新改成大面窗戶後，L型轉角採光面能引入完整的光線，視覺感也更加俐落。

自然米色調營造放鬆氛圍

主臥延續整體空間簡約調性,以自然的米白色調加上床頭天花板間接照明,營造溫暖放鬆的寢居氛圍;主臥衛浴同樣採用米色調磁磚,使整體感更為協調一致。

CASE 3

HOME DATA

屋齡｜約 31～40 年　屋型｜電梯大廈　坪數｜39 坪　家庭成員｜夫妻
建材｜乳膠漆、超耐磨木地板、灰鏡、木作貼皮、烤漆、鋼刷木皮、系統板材、大理石、磁磚、木百葉、賽麗石、石材、噴漆

大幅挪移客浴，
打造寬闊無障礙退休宅

空間設計暨圖片提供｜賀澤設計　文｜陳佳歆

●移除臥房和客浴之後，以開放式規劃展開公領域空間的尺度，解決原先光線不足問題，空間感更為輕盈明亮，原本客浴的管道間則不著痕跡地隱藏在櫃體裡。

Before　　　　　　　　　　　After

　　一對屆臨退休的夫妻想重新改造居住了十幾年的老宅，除了覺得原本格局已經不符合目前生活需求，一方面雖然目前身體還很硬朗，仍希望提前規劃適合年老的居住退休宅。

　　平時有下廚習慣的女主人，覺得廚房太小不好使用，而且居中的餐廳採光也被其他臥房阻檔，加上平常只有夫妻倆人居住，有些臥房閒置已久，整體空間顯得很昏暗沒有朝氣。

　　從無障礙空間的概念為基礎來規劃這間老宅，首先整平玄關和公領域地坪的些微高低差，並且沿窗邊增加臥榻，方便作為進出門穿脫鞋使用，公領域則果決地移除鄰近廚房的次臥，同時大幅度將客浴移到另一側主臥衛浴旁，讓出完整的空間給客廳、餐廳和廚房，形成一個開闊的公領域，不但打開了廚房尺度也為中間引入更充足明亮的光線。

　　主臥衛浴考量老年使用安全，採用無門檻設計同時加大入口門寬，盥洗檯面則延續夫妻倆多年使用習慣，採用分離式規劃配置在衛浴外，為解決大幅移動客浴造成地板墊高的狀況，將其中一間馬桶改採用壁掛式，使得安全性和整體美感都能兼顧。

　　改造後的空間以暖白搭配天然木色，襯托出簡約而不失溫馨的居家感，讓家有截然不同的生活樣貌，原本不常邀請親友來訪的夫妻倆，在裝潢完成後人際交流漸漸頻繁，對於往後老年生活有更多不同的想像和期待。

整平地板落差降低絆倒意外

處理原本玄關和客廳地坪間的高低差，減少年老步伐不穩可能絆倒的情況，同時沿著窗邊規劃臥榻作為穿鞋椅使用，也成為一處可臨窗休憩的位置。

擴增廚房打造理想下廚環境

原本的次臥與廚房整併，重新配置 L 型廚房檯面和中島，讓平時習慣在家料理餐食的女主人，有更寬敞的備餐空間可以使用，也提升人與空間的互動性。

思考細節賦予多元空間功能

多功能室採用活動拉門，使空間更能依照需求靈活運用，一旁的客用浴門檻特別做了 45 度導角門檻，進出也不會有絆倒的顧慮。

無障礙設計提升衛浴使用安全

考量夫妻倆未來老年生活，主臥衛浴加大門寬至 100 公分並且採用無門檻設計，盥洗檯則沿續原本使用習慣規劃在浴廁外面方便使用。

老屋裝修基礎課

CASE 4

HOME DATA

屋齡｜約 40 年　屋型｜公寓　坪數｜75 坪　家庭成員｜夫妻、2 小孩
建材｜海島型實木地板、PERGO 木地板、進口磁磚、木皮板、塗料、烤漆、FRP 板、鐵件

旋轉美梯煥新 40 老屋陳舊感

空間設計暨圖片提供｜森參設計　文｜Fran Cheng

● 從鋼琴區、開放餐廚區、客廳到玻璃書房的無間斷景深，搭配多面向的採光，完全改變原本多隔間且封閉的格局，再搭配每個角落都能看見的美型旋轉梯更顯華宅優雅。

Before　　　　　　　　　　　After

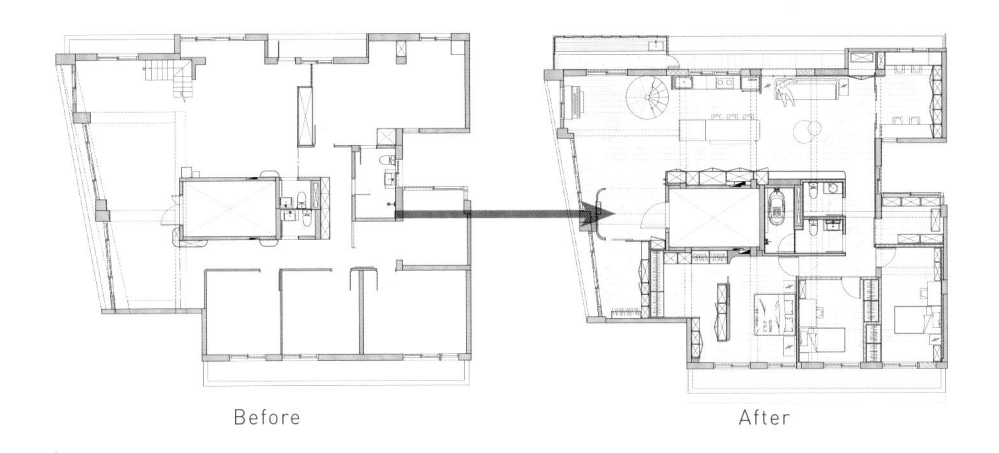

　　原屋主買兩戶公寓打通後，再加上頂樓共有75坪，坪數雖大，但因內部房間多導致採光差，加上老式樓梯使氛圍沉重、陰暗。

　　但育有二個小孩的現任屋主希望有寬敞、可跑跳的明亮大空間，所以設計師先大刀闊斧地砍掉隔間牆重建格局。

　　首先，玄關以白色造型牆搭配可透視的屏風引入後方窗光，左側配置大衣帽間，右轉進入大廳前則利用窗邊畸零區規劃木榻區，日後可作為鋼琴區。

　　接著將舊型L型樓梯拆除，請結構師計算重建輕盈旋轉梯，不僅優雅、吸睛，也讓開放餐廚區與客廳景深拉長變大；至於客廳旁則以玻璃拉門串聯書房，看書、活動都更自由寬敞。

　　私密區選擇在同一區保留三間房，除了以回字動線的更衣區擴大主臥格局，二間小孩房均以機能滿足為主，頂樓鋪以木質臥榻作為遊戲室或客房，至於收納設計除有儲藏室、衣帽間，餐櫃及客廳沙發後等畸零小空間也都成為最佳收納櫃。

白格透光屏風大器迎賓

在玄關落塵區先以圓弧玄關牆搭配格狀透光白格子屏風營造端景，讓賓客一進門就能感受寬敞大宅的明快與器度，而左側木門一開則是走入式衣帽間，配置鞋櫃與大型櫥櫃，相當大方。

旋轉梯與鋼琴區增添光感

傳統 L 型樓梯讓空間感覺呆板又沉重，討論過後決定請結構技師重新計算承重標準，設計白色旋轉梯成為入門後第一焦點；而右側臨窗的區域則留白並鋪上木地板作為舞台，讓公領域更明亮外，未來會擺放鋼琴。

玻璃背牆為客廳引入光源

為改善採光差的問題，客廳沙發背牆局部採用玻璃格子窗來引入矮窗光源，另一側則銜接木格子牆整合後方畸零區，並規劃成儲物櫃。考量老公寓天花板較低，大樑也僅以導圓弧線裝飾，呈現簡約之美。

玻璃書房內有滿滿書櫃

一家四口不需要過多房間，所以撤掉臥室隔間牆，改為玻璃格子拉門書房，書房除了兩側均設書桌，側牆則有滿牆書櫃，不僅可放大量書物，也可讓一家四口在這裡看書、處理雜務。

CASE 5

HOME DATA

屋齡｜約44年　屋型｜華廈　坪數｜32.5坪　家庭成員｜夫妻、1幼兒
建材｜系統櫃、天然木皮、訂製鐵件、復古磚、超耐磨地板

打好老屋基底，
迎接未來各種生活可能性

空間設計暨圖片提供｜孛沃設計　文｜喃喃

● 在格局不做更動前提下，採用鐵件玻璃門片來保留空間彈性，並以纖細的金屬門框與分割比例設計增添設計感，門框則烤漆上沉穩的藍，來注入些許活潑氣息，同時呼應整體空間氛圍。

Before & After

　　原本是租屋一族的屋主，買房條件很簡單，一是希望不要離兩人工作的地方太遠，再來是有三間臥房需求，但周遭一直沒有新房子釋出，於是在符合地理位置及房間數量兩個條件下，他們選擇了這棟屋齡約有四十幾年的老房子。

　　四十幾年的老屋，雖然大樓社區有做好維護，但老屋該有的漏水、管線問題仍不可避免，因此不只基礎工程幾乎全部重作，更因為位於頂樓，漏水問題及防水工程也特別慎重處理。

　　除此之外，由於屋形是不規則的扇形，若以櫃體來拉直空間線條，會讓空間變得狹小，也少了使用彈性，所以除了規劃必要的收納櫃體外，其餘則以活動家具來適應原始屋形，空間的運用更靈活，少見的扇形也意外成為吸睛亮點。

　　由於屋主的小孩還小，生活型態勢必在未來幾年會有所改變，因此格局不做變動，而是以軟裝、家具家飾及漆色等方式，來賦予老空間新樣貌，如此一來既能展現屋主的居家美感與品味，同時也能保留未來生活可能性。

選用活動家具讓空間使用更自由

為了因應未來可能改變的生活型態，保留原始格局，以活動家具來定義空間屬性，原本做為客廳使用，未來也許會成為放置小朋友玩具的玩耍空間，隨著當下生活模式，空間的使用方式隨時可以變化。

既能界定空間也是視覺亮點

在餐廳入口位置，設計了一個古典的拱門造型，藉此有呼應空間的弧形元素目的，並製造出空間視覺焦點，同時也能在以開放格局為主的空間裡，做出明確的區域界定。

通透材質引入光源,增添空間明亮感

設定為工作室的區域,是空間裡唯一沒有採光的位置,因此採用清透的玻璃材質來做為隔牆,不只可讓視線穿透而保有空間開闊感,同時也能引入客廳光源,讓工作室不至於太過昏暗。

簡約卻不失溫馨更好眠

臥房本來採光就不錯,因此收納櫃即便採用略顯沉穩的木質調,也不會讓人感到陰暗,反而可以調和過白的空間,讓人情緒沉澱下來,感受能安心入眠的睡寢氛圍。

CASE 6

HOME DATA

屋齡｜約 50 年　屋型｜公寓　坪數｜35 坪　家庭成員｜夫妻、2 小孩
建材｜磁性板、磁磚、木地板、長虹玻璃

重整結構與漏水、機能升級，50 年老屋迎來新風貌

空間設計暨圖片提供｜十一日晴設計　文｜Celine

● 透過格局調整，老屋空間更明亮舒適，煙燻木色與灰白色調相互呼應，營造沉穩而富有層次的美式復古風。細緻的收納設計與開放式格局提升生活機能，讓舊宅蛻變實用與美感的理想居所。

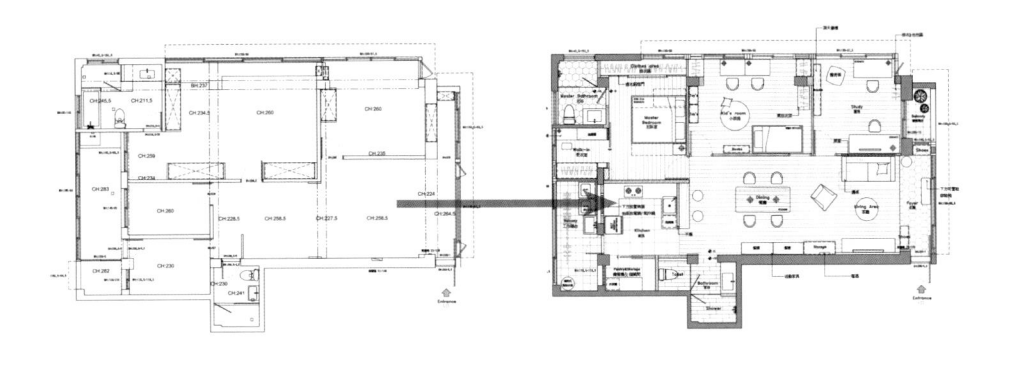

Before　　　　　　　　　　　　After

　　這間位於台北市中心、屋齡超過半世紀的老屋，雖然地點優越、坪數寬敞，卻因歲月侵蝕產生結構與生活機能問題，包括漏水、壁癌、收納不足，使居住品質受影響，屋主因此決定全面翻修，讓老屋重獲新生。

　　改造首要任務是解決防水問題，包含頂樓與外牆漏水，以及拆除天花板後發現鋼筋膨脹、粉刷層掉落，須進行防鏽與水泥回填處理，以確保結構安全。另外，老屋地板高低落差達 5～6 公分，需拆除見底後重新整平再鋪設新材質。

　　格局部分，原本過大的主臥縮減面積，多出的空間規劃為小孩房與遊戲室，並預留雙門與隔間設計，未來可彈性調整為兩間獨立房間；廚房則改為半開放設計，提升光線與空氣流通性，並增設儲藏間，解決收納問題。

　　屋主偏好美式復古風，特別喜愛煙燻色木質調，因此選用深色木材搭配白色與灰色地磚，並融入黑鐵與古銅等工業風元素，讓空間更具層次與個性。

　　此外，書房局部恢復外推陽台，打造景觀綠意角落，為偶爾需要遠距工作的屋主提供放鬆氛圍。

　　這次改造不僅提升安全性與機能，更透過風格整合，讓老屋蛻變為兼具實用與美感的理想居所。

開放書房，串聯溫暖日常

客廳與書房以開放式設計串聯，讓光線與空間感延伸，木作造型窗框設計增添視覺的豐富與層次，同時保留通透感，使閱讀與休憩空間更具靈活性。

開放式廚房，增進互動提升明亮感

透過開放式設計，讓廚房與用餐區更顯寬敞通透。一旁巧妙規劃獨立儲物間，既可收納家電還能放置過季、清掃家電，滿足屋主對於空間美感與實用性的需求，使日常料理時光更加愉悅。

掛衣區結合更衣間，
滿足大量衣物收納

主臥室融合美式現代風格，鋁框拉門烤棕色調，與公領域顏色更為融合協調，長虹玻璃材質讓視覺具通透性，另一端則配置為走入式更衣間，提供豐富充裕的衣物收納與分類使用。

穀倉門五金創造復古氛圍浴櫃

客浴利用原始畸零結構，規劃出三件式衛浴設計，雙色磁磚配色，加上木作底框、鋁框門、穀倉門五金裁切搭配運用，創造出工業復古感的獨特浴櫃。

CASE 7

HOME DATA

屋齡｜約 46 年　屋型｜公寓　坪數｜33 坪　家庭成員｜夫妻、1 小孩
建材｜超耐磨木地板、炭化樺木、塗料

微整 33 坪老屋，廚衛調度更靈活

空間設計暨圖片提供｜福研設計　文｜EVA

● 由於客廳有著西曬和漏水問題，因此牆面重新施作防水，解決漏水問題，表面特意鋪陳洞洞板，賦予收納機能。而洞洞板與牆面之間也形成空氣層，能強化隔熱效果，內部也能隱藏管線，營造簡約俐落的空間視覺。

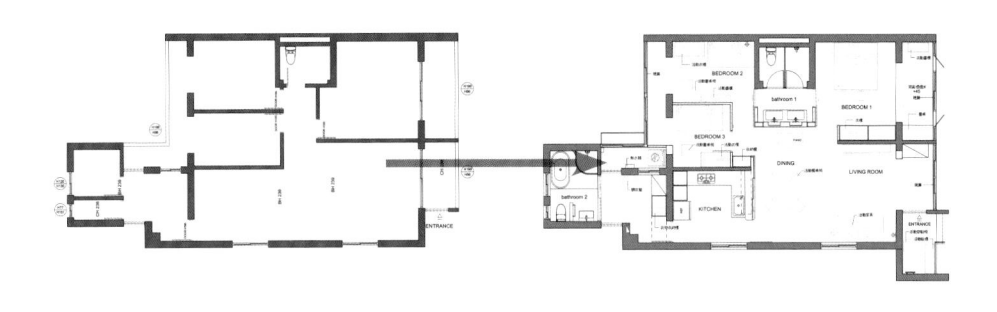

Before　　　　　　　　　　　　After

　　這間位於民生社區的老屋是屋主夫妻的老家，雖然有著良好採光，但原始格局不符需求，老屋漏水也影響居住品質。因此期待翻新後能擁有動線順暢，富含彈性的居住環境，和學齡前的孩子一同成長。

　　考量到座落於邊間，沙發一側有西曬又有漏水問題，因此重新施作防水後，改以洞洞板鋪陳背牆，不僅擴增收納空間，強化隔熱效果，日後也方便拆卸，防水維護更便利。

　　而原有的廚房向外挪移，擴大使用空間，同時搭配拉窗和門片，形塑獨立又開放的空間，通透的視野方便隨時看顧孩子。

　　至於臥寢區則維持靈活運用空間的初衷，將衛浴改為乾濕分離，洗手台外移，並串聯主臥與次臥動線，有效提升盥洗效率，也增加親子的親密互動。

　　兩間次臥的隔牆向後退縮，保有共用陽台，形塑能自在遊走兩房的回字動線。而位處角落的客衛則重新整合，納入屋主夢想的浴缸，完善洗浴機能。一旁原有的廚房則改為洗衣房使用，打造洗浴、晾衣一氣呵成的家務空間。

墨綠色調呼應戶外綠意

玄關拆除拉門,與客廳相連,形塑通透開敞的視野。融入女屋主喜愛的墨綠色調,巧妙與戶外綠意呼應,同時輔以弧形曲線修飾樑體,空間線條更為柔和。客廳中央則安排木樑貫穿,不僅藏入投影布幕,也暗藏燈光,增添溫潤暖意。

彈性拉窗與門片,廚房運用更靈活

原始廚房相對封閉狹小,將廚房外移改為開放,客廳與餐廚形成一體,空間開闊顯大。特地運用洞洞板設計廚房門片,再搭配拉窗,需要時能形塑獨立的廚房空間,有效避免油煙影響,也顧及孩童安全。

主臥增設臥榻，機能更多元

主臥以英倫藍鋪陳，奠定寧靜平和的臥寢氛圍。同時沿窗安排臥榻，不僅多了放鬆悠閒的空間，也善用畸零角落設置層板，臥榻下方也多了收納，滿足多元機能。

分離式衛浴，串聯主次臥動線

為了有效利用空間，中央狹小衛浴改造為分離式，馬桶、淋浴間獨立設置，洗手台外移，搭配雙面盆設計，方便家人共同使用。衛浴兩側安排入口，同時與主臥、小孩房相連，打造能自由行走的連通房，照顧小孩更便利。

CASE 8

HOME DATA

屋齡｜約20年　屋型｜華廈　坪數｜28坪　家庭成員｜夫妻、1小孩
建材｜特殊塗料、玻璃、木地板、系統板材、長虹玻璃、鐵件

舊宅重生，機能與美感並存

空間設計暨圖片提供｜寬月設計　文｜Celine

◉ 以奶茶色牆面與清水模塗料打造溫暖氛圍，圓弧形天花柔化大樑線條。玄關設置大型鞋櫃與隱藏式收納，讓空間功能性與美觀兼具，動線更加流暢。

Before → After

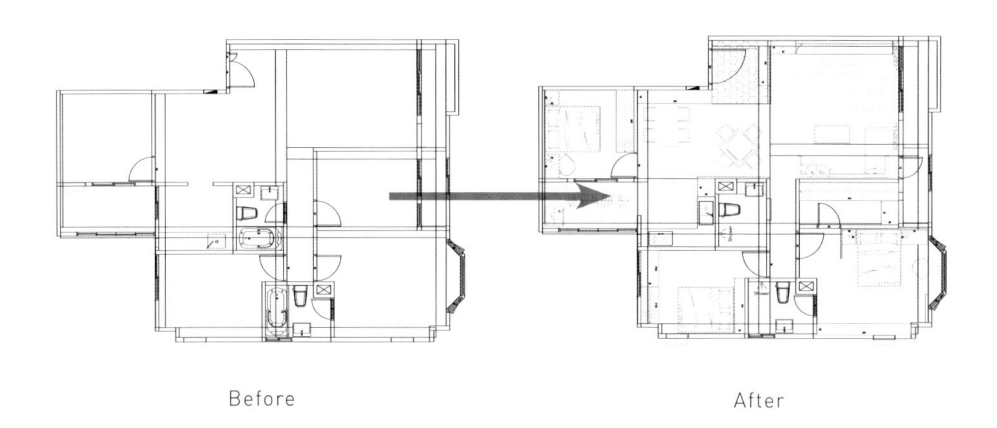

每一間老屋都有一段故事，而這次的改造，則從窗邊的漏水問題開始，延伸到全屋空間的重新規劃，設計師以「回歸原始，保持純粹」的理念，藉由現代溫馨的設計手法，賦予老屋嶄新的生命力。

改造的第一步是解決結構問題，拆除老屋窗戶重新施作防水、更換新窗，確保居住安全與舒適性，同時，將四房改為三房，釋放出的空間給予客廳與主臥更衣室，提升兩個場域的使用機能。

客廳是改造的核心亮點，牆面採用奶茶色與清水模塗料相結合，既溫暖又個性十足；圓弧形天花從入口延伸至廊道，不僅增添柔和流暢的線條美感，還巧妙修飾了視覺突兀的樑柱。

始進門即見客廳的格局，設計師新增大型鞋櫃，劃出功能完善的獨立玄關，讓空間更有層次感。廚房局部拆牆後改為開放式設計，並配置 L 型廚具與完善的電器櫃，大幅提升使用效率。

緊鄰廚房的餐廳則配有多功能餐櫃，融合玻璃吊櫃、層板與半腰櫃，兼具展示與實用的收納需求，呈現美觀與功能並重的用餐場域。

每個房間都融入巧思：主臥的角窗搭配木皮造型，修飾樑柱的同時提升視覺層次；床頭平台則結合柔和光源，營造出溫馨的睡眠氛圍。小孩房則善用床頭空間，打造儲物功能，卽使空間不大，也能具備完整的使用機能。

多功能餐櫃提供豐富收納

餐廳配置多功能餐櫃,結合玻璃吊櫃、層板與半腰櫃,滿足展示與多樣化收納需求。搭配溫潤的材質與精緻設計,營造實用且富有品味的用餐空間。

開放廚房優化動線與採光

廚房設計採開放式規劃,搭配 L 型廚具與完整電器櫃,充分提升空間利用率與使用便利性。流暢的動線結合明亮材質,打造功能性與視覺效果兼備的理想料理場域。

CHAPTER 3 空間實例

主臥擴增更衣提升實用性

主臥床利用長虹玻璃隔屏設計，避免直視床鋪，並利用拆除一房後的空間部分規劃為更衣間。此外，主臥角窗以木皮造型包覆樑柱，增添視覺層次感；床頭結合平台設計與柔和光源，營造舒適氛圍。

善用床頭創造收納機能

小孩房充分利用床頭空間提高收納，同時搭配溫暖的材質與簡潔的設計，即使空間不大，也能滿足成長需求，營造舒適實用的小天地。

CASE 9

HOME DATA

屋齡｜約 26 年　屋型｜公寓　坪數｜32 坪　家庭成員｜夫妻、1 小孩
建材：PERGO 木地板、進口磁磚、木皮板、司曼特、烤漆、鐵件、人造石、伊諾華板

挪一房換書牆大廳變寬變美

空間設計暨圖片提供｜森叄設計　文｜Fran Cheng

● 經過格局微調後，利用沙發後方樑下空間規劃前後雙排的書櫃來收納屋主藏書，書櫃搭配滑軌使用起來更方便，而右側粉色門櫃不僅能收納不規則雜物，也有助於提高整體牆面顏值。

Before

After

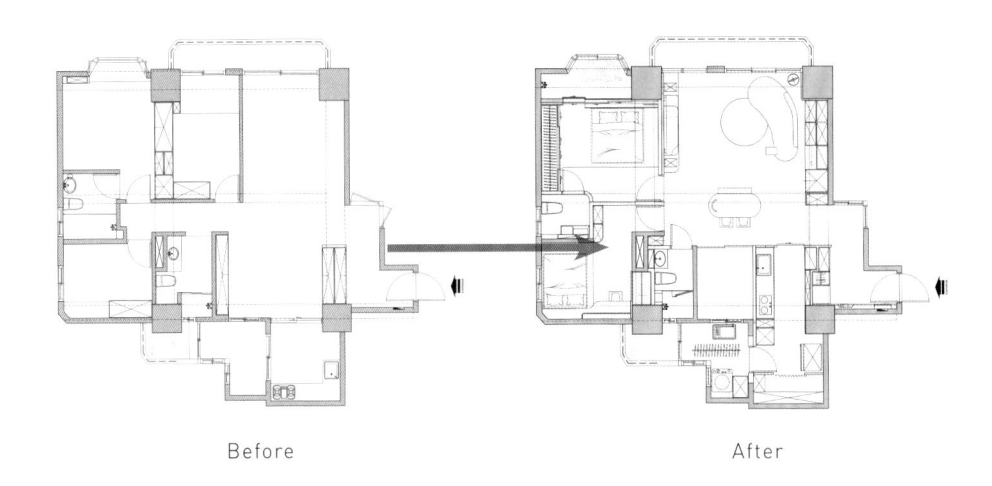

　　屋主是醫生，所以對於裝修建材特別重視，除了強調要無毒健康外、也要求要 MIT。

　　至於在格局上主要考量原廚房位置不好用，餐廳空間占比也過多，為了提高坪效，決定將客廳旁的小房間拆除來放大公共區，這樣就能讓電視牆與沙發往前移，且可以利用沙發後方樑柱下的畸零空間規劃一座前後雙排櫃體，再搭配滑軌設計以便收納屋主大量書籍。

　　另一方面，將原本餐廚區改成玻璃隔間的半開放書房與廚房，廚房後方還能多出一片儲藏間，而位於動線上的餐桌區也能兼作家人共讀、聊天的空間。

　　原本門對門的主臥室與次臥房，同時將門向都移位轉向餐區，接著利用隔柵木飾板拼接特殊灰泥漆塗料牆，打造餐區的隱形門裝飾牆，也連結電視主牆，有助於放大公共區整體視覺。在主臥中因格局調整得以打造出私人陽台，並將浴缸移至陽台，除可增加更多收納空間，住起來也更顯寬適有餘。

玄關雙面櫃區隔出落塵區

玄關格局動線寬敞，除了規劃污衣櫃、衣帽櫃與鞋櫃外，搭配地板就區隔出落塵區，同時在側牆則以洞洞板與造型鏡來增加穿衣整裝與吊掛功能，至於玄關櫃後方設計有廚房的吧檯櫃。

移走一房，瞬變明亮大格局

原本客廳與主臥中間的小房間拆除後可放大公共區格局，同時也因少了隔間牆而使採光窗串連變得更明亮，搭配矮背沙發與後方成排書櫃可滿足更多收納需求；另一方面，主臥也可因此而略為放大。

電視主牆無痕隱入兩扇房門

將電視牆後的主臥與次臥門片轉向對著公共區，接著以隔柵木飾板搭配特殊灰泥塗料來拼接出面寬更寬敞的裝飾主牆，且將兩扇臥室門隱形設計融入其中。

重塑格局打造私人沐浴陽台

主臥轉了門向後讓床位更好安排，同時將床頭略為內縮，以便在床頭砌牆打造出私人陽台，接著將浴缸移至私人陽台，除了可享受充足的天光，還配搭有木百葉窗來確保隱私性，同時能讓主臥衛浴間更寬敞。

CASE 10

HOME DATA

屋齡｜約40年　屋型｜公寓　坪數｜24坪　家庭成員｜夫妻
建材｜磨石子磁磚、玻璃、塗料

融入藤編、海棠花玻璃與綠色，營造自然沉穩復古風

空間設計暨圖片提供｜日居設計　文｜Celine

● 原本配置於空間深處的餐廳挪移至客廳旁，以餐桌為家的核心創造出環繞式生活動線，復古藤編系統櫃不但整合多元收納機能，也隱藏起電箱，同時回應屋主喜愛的復古氛圍。

Before → After

　　這間位於巷弄內、屋齡約四十年的老公寓，原本的採光與通風條件不佳，室內還存在壁癌與地面不平等問題，設計團隊根據屋主需求，從功能與美感兩方面進行全面改造。

　　將老屋窗框拆除重新施作防水、地面磁磚也全部剔除重新鋪設木地板，避免踩踏可能產生的凹陷。

　　再來是打破傳統空間配置，將客廳、廚房、中島餐廳串聯，讓家人互動更自然密切，再加上客廳捨棄沙發與電視模式，改以長餐桌作為家的核心，不僅能用餐，也能成為閱讀、聚會等多功能區域。

　　至於老屋無法避免的單向採光問題，則透過廚房的玻璃拉門，加上後陽台採用三合一通風門扇，讓光線可以從客廳延伸到空間深處，提升室內明亮與通風效果。

　　不僅如此，原本老公寓進門直通客廳、缺乏玄關的格局，也經過設計團隊重新調整，將局部客廳尺度讓出規劃為玄關，並利用一道矮牆、磨石子地磚區隔出落塵區，更兼具屋主所嚮往的園藝植栽區。

　　色彩搭配上則運用白色與原木色為主調，並調入深綠色作點綴，融合男女主人偏好的北歐與復古風，既溫暖又不失沉穩，透過改造充分展現老屋翻修可能性，不僅讓空間實用性倍增，也為屋主打造兼具風格與功能的理想之家。

融入自然綠意的玄關設計

面對老公寓進門後無玄關的格局問題，設計團隊巧妙釋放出局部客廳區域，並以磨石子磁磚地板和矮牆劃分出玄關空間。這裡不僅是玄關，同時也將屋主喜愛的植栽園藝融入其中，為家帶來自然清新氛圍，每次進門即可感受充滿綠意盎然的舒適感

溫暖沉穩的復古調性

為平衡屋主倆人對喜愛的風格差異，整體空間以白色和木色為基調，並巧妙融入綠色立面，加上深木家具搭配，包括壁燈也是海棠花玻璃製作而成，呈現出和諧、復古沉穩的氛圍。

玻璃拉門提升光線、串聯互動

廚房採用玻璃拉門為隔間，結合後陽台三合一通風門設計，讓光線可以從客廳區域向內延伸，有效解決老屋後段空間採光不足的問題，同時也提升空間的開放性，帶來更緊密的互動。

雙色塗料與軟裝勾勒生活感

主臥捨棄多餘櫃體設計，牆面利用雙色塗料的層次設計，搭配藤編家具與植栽點綴，營造自然溫暖的生活角落。

CASE
(11)

HOME DATA

屋齡｜約16～30年　屋型｜公寓　坪數｜27坪　家庭成員｜1人
建材｜油漆、超耐磨木地板、鐵件、烤漆、系統板材、人造石、軌道燈

少一房收整廊道，
重塑老公寓的當代風貌

空間設計暨圖片提供｜賀澤設計　文｜陳佳歆

● 原本一進門就是冗長的廊道，在少了一間臥室後減少昏暗的空間感，公領域更明朗開闊，充滿綠意的山景也從底端被隱約帶入室內。

Before　　　　　　　　　　　　　　　After

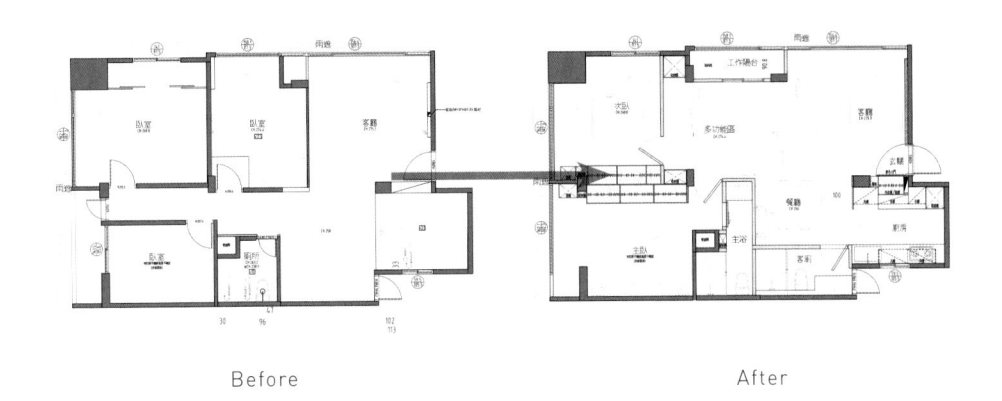

　　父母為北上工作的兒子購入了這間老公寓，早期 27 坪三房的格局令人感覺封閉，一進門就直對昏暗的長長廊道，外牆和衛浴也有老屋常見的漏水問題，而且只有一間衛浴也沒有乾濕分離，空間雖然老舊，但位置鄰近山邊，採光也還算不錯，便希望能逐一突破老屋現況，依照生活所需重新調整佈局。

　　整理老舊管線與漏水問題後，接著大幅重整空間格局，由於平時只有一人居住，父母偶爾來探訪，於是將靠近客廳的次臥移除，使公領域感覺更為開闊，開放空間作為練琴及健身使用。

　　而原本狹長的廊道空間被收進底端的兩間臥室，走道變短後不只減少冗長與昏暗感受，臥室空間也變大，同時為臨近山景的主臥開一扇大窗，早晨只要拉開窗簾便能納入更多濃郁綠意，並且在主臥衛浴之外，另外增加一間客浴以符合使用需求。

　　雖然空間以墨黑及暖灰鋪陳，開展格局後在日光的陪襯下不覺沉重，反而多了分理性的沉著，另外為父母使用的次臥帶入較明亮的顏色，呈現不同的空間風情。

日光與墨黑色交織空間個性

回應男主人個性和喜歡的色調,空間以濃郁的墨黑色為主鋪陳,局部搭配暖灰,在展開公領域空間後的明亮日光中,擁有更豐富的空間層次。

收入走廊空間放大臥房舒適度

原本狹小的臥室,在收進走廊及後陽台空間後放大許多,調整為主臥套房,再利用大幅開窗框起臨近山邊的綠意,營造出放鬆舒適的寢居氛圍。

因應長輩習慣轉換次臥色調

考量父母對深色接受度不高，次臥沒有帶入主色調的黑，置入明亮的白色櫃體，並搭配溫暖的自然色調，在個人喜好和長輩感受之間取得平衡。

打開公領域讓空間更有生活感

移除一間位在客廳旁的次臥，作為多功能區滿足男主人在家練琴及健身的需求，同時帶來開闊明亮的公領域，一掃老屋侷促感，提升居住品質。

CASE 12

HOME DATA

屋齡｜約 50 年　屋型｜電梯大廈　坪數｜35 坪　家庭成員｜3 大人、2 小孩
建材｜磁磚、超耐磨木地板、實木貼皮、水磨石

拆臥室、不做客廳，50 年老屋放大兩倍空間感

空間設計暨圖片提供｜隹設計　文｜EVA

● 餐廳安排木質長桌，一旁則銜接高櫃，巧妙取代傳統客廳，成為全家用餐、工作或閱讀的生活重心。空間色調以木質、深綠和灰階磁磚交織，營造穩重溫暖的氛圍。

Before　　　　　　　　　　　　After

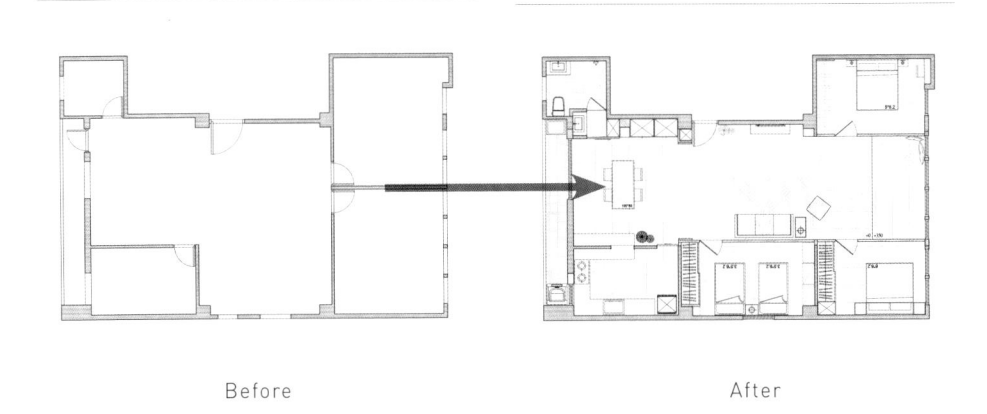

　　這間五十年的老屋是屋主自小生長的老家,歷經家庭成員的變動,兩房格局已不符需求,光線被臥室阻隔,僅剩單面採光,空間也狹窄陰暗且老舊。因此屋主期待能重塑空間,改善採光、動線與隔音問題,與家人共享舒適開闊的居住環境。

　　由於客廳光線不足,臨路大窗老舊帶來噪音干擾,再加上屋主夫妻、兩名女兒與屋主姐姐一同居住,要有充足的臥室,因此率先位移臥室,並增設至三房,滿足臥寢需求。

　　拆除臨窗臥室後,有效引入大量採光,同時更換窗戶,解決隔音問題,形塑明亮寧靜的氛圍。沿窗設置臥榻,客廳、餐廳也安排在同一軸線上,公領域更通透。去除傳統的客廳意象,特地不放沙發、不擺電視,客廳與臥榻相連,席地而坐的設計打造一同閱讀放鬆的親密場域。

　　由於屋主以烘焙爲業,廚房採用U字型設計,擴展料理檯面,賦予開闊寬敞的空間質感。與餐廳之間更留出送餐口,搭配內外吧台的設計,不論是當作早餐吧台或送餐台,機能更多元完善。

拆除隔間，公領域放大空間感

拆除臥室，引入自然採光，改善原有的陰暗格局。同時客廳、餐廳也安排在同一軸線上，串聯開闊通透的公領域。客廳不放沙發、電視牆，僅沿窗設置臥榻，隨時都能席地而坐，打造悠閒放鬆的氛圍。

擴增 U 型台面，廚房更大更實用

由於屋主從事烘焙業，對廚房有一定的堅持，安排 U 型廚具擴增台面空間，不論在哪兒備料都游刃有餘。為了節省來回餐廳的力氣，廚房牆面特意留出拉窗，增設吧台，形塑快速送餐的動線，也維持通透開敞的視覺。

雙洗手台，提升盥洗效率

考量到家庭人員眾多，在僅有一間衛浴的情況下，增設洗手台讓多人能同時盥洗使用，提升生活效率。而繽紛亮麗的水磨石台面則賦予空間活力，注入清新氣息。

老屋裝修基礎課

CASE
(13)

HOME DATA

屋齡｜約47年　屋型｜電梯華廈　坪數｜37坪　家庭成員｜1人
建材｜天然木皮、實木地板、特殊漆、珪藻土、訂製鐵件、大理石

結合天然材質與無障礙設計，
不只活化老宅住得更安心

空間設計暨圖片提供｜季沃設計　文｜喃喃

● 空間裡雖大量使用木素材，但藉由交錯使用不同木紋、木色，及木格柵設計，來讓同一種材質呈現出和諧的視覺效果與變化，另外輔以白色搭配，達到提亮空間目的，同時也注入簡約俐落感。

CHAPTER 3 空間實例

Before → After

　　老屋屋齡已有四十幾年，不只要進行基礎工程維修，更因為經過好幾手改造裝修，格局變動太多，因此設計師特別調出原始平面圖，來與現有格局做對照，除了可以更好掌握接下來的格局配置，也有利於工程進行時的施工判斷。

　　房子未來只有屋主一人居住，因此將原來的三房格局調整成兩房，並藉由把鄰近廚房的兩房整合成一房，來拉直空間線條，避免產生難以利用的畸零地，廚房位置則順勢稍微挪移，從新佈局成開放式廚房，如此一來便可與餐廳、客廳、書房等公領域進行串聯互動，整體空間也變得寬敞且開闊。

　　除了格局的調整，考量未來年紀漸長，趁著這次翻修，在空間裡加入無障礙設計，並將過往因廁所移位而產生的高低差，全部調回水平，未來不只行動無障礙，行走於空間也更自在且安全。

　　除此之外，大量採用木素材、石材等天然建材，利用材質的原始質感，來營造不會過度精緻，簡約寧靜的生活氛圍。

經過調整挪移，
生活動線變得合理且舒適

廚房挪移後變小且較為方正，除了基本廚具配置外，往餐廳方向規劃一座中島，電器櫃與中島位置相鄰，接著再從中島延伸出餐桌，如此一來料理空間變大，動線也更順暢，若有朋友來訪，則成為大家聚集互動的區域。

推拉門片設計讓空間
更靈活好用

推拉門是無障礙空間常用的設計，在這裡，推拉門採取木格柵設計來淡化門片存在感，同時亦巧妙成為背牆設計元素之一，而打開拉門裡面是書房，平時可依使用情境選擇打開或關上，使用相當彈性

採用自然材質注入溫暖調性

透過格局重新規劃，主臥移到距門口較遠的位置，刻意與公領域做出劃分，讓睡寢空間可享有隱私性，風格上則注入屋主喜歡的自然元素，以木素材為主，再加入大地色系及手感漆面點綴，營造出放鬆氣氛，相當有助於入眠。

玻璃材質引入光線打亮空間

主臥衛浴採乾濕獨立分離規劃，乾區缺少採光，加上使用頻率較高，因此採用玻璃推拉門，利用玻璃材質引入光線，增添空間明亮感，而沒有門檻的設計則能消除斷差，行走沒有障礙也更安全。

老屋裝修基礎課

CASE 14

HOME DATA

屋齡｜約37年　屋型｜公寓　坪數｜32坪　家庭成員｜夫妻、3小孩
建材｜木皮板、科品門、進口磁磚、烤漆、玻璃、白板、鐵件

甜漾木白．和菓子居家

空間設計暨圖片提供｜森叁設計　文｜Fran Cheng

●屋主因為喜歡日式氛圍，所以除了建材上運用大量木質與白色牆，也將天花拉高、採用穿透木窗隔間來引入更多光源，展現宜居舒適的無印風格。

Before　　　　　　　　　After

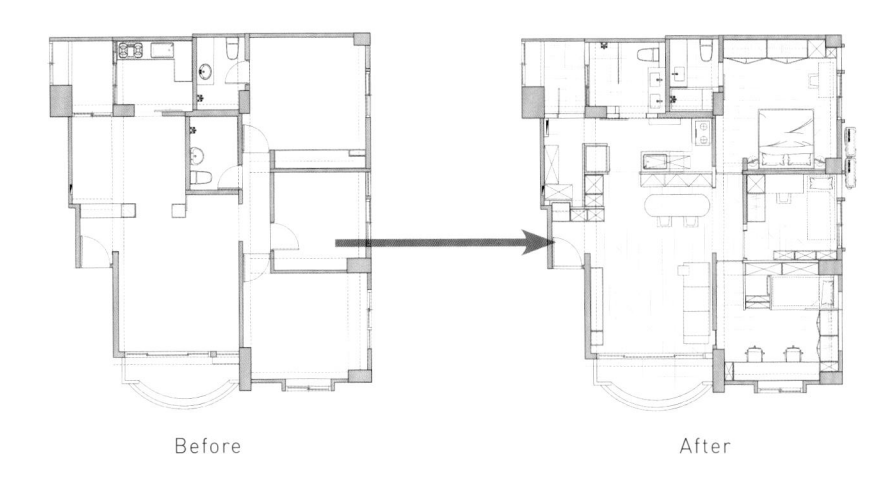

　　屋主夫妻與三個兒女想搬進這 32 坪的老屋，第一個要挑戰的是只有三房二廳一家五口如何住？

　　再來則是舊有裝潢做了間接光源而壓低天花板，採光暗且收納也不足，面對這些問題，設計師決定從坪效差的玄關與餐、廚房開始調整。

　　首先，將舊有一字型玄關端景櫃改成 L 型牆櫃，既滿足玄關與餐廚區的收納需求，還能在舊餐廳位置增設一間儲物間。

　　另外，將廚房外移與客用衛浴間對調後，讓餐廳與廚房之間可藉由木格子窗作連結，半開放的廚房是屋主喜歡的日式風格餐廳，還可迎來明快光感，同時也讓一直卡在房子中央的客浴有更好歸處。

　　客廳以白與淺灰色為主軸，拿掉間接照明改以軌道燈可拉升屋高、減少些壓迫感，搭配座榻式餐區展現輕鬆氛圍。

　　因為只有三間臥房，除主臥外，最小房間作為男孩房，較大房間則以上下舖讓兩位女兒同住，搭配雙人書桌、雙倍衣櫥，滿足公主們的需求。

L 型牆櫃提升多區收納力

左側入門區以 L 型收納櫃包覆轉角，既能滿足玄關收納、也可以做為餐區電器櫃及廚房冰箱區，同時冰箱後方則增設一間儲藏室，可更有系統性地收納大型物品。

拉升屋高換嵌燈更通透明快

舊裝潢因為間接光源壓低天花板，也影響採光，拆除後改以嵌燈與軌道燈，讓天花板拉高且更簡潔，另一方面，鄰接小孩房的牆面加裝一道玻璃窗展現通透感。

CHAPTER 3 空間實例

移客浴改餐廳換得流暢動線

原本進入主臥旁有客用衛浴卡住，導致動線與房門位置受限，翻新後將客浴移置廚房後方，再搭配開放餐廚設計，除了能讓公共區空間放大且更流暢外，主臥房門也能向外延伸些。

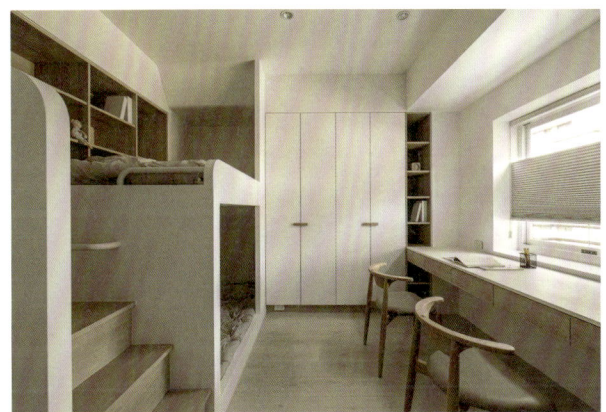

雙人包廂房功能齊全又舒適

因為不想變更三房格局限縮公共區，或導致房間過小難以居住，決定讓兩位女兒共住較大房間，並以包廂式上下舖，搭配雙桌、雙櫥、雙格子櫃等設計，空間開闊又能展現超高功能性。

CASE 15

HOME DATA

屋齡｜約30年　屋型｜公寓　坪數｜30坪　家庭成員｜2大人
建材｜油漆、胡桃木皮、大理石紋地磚、紅陶磚

挪移主臥和衛浴，打造雙主臥格局

空間設計暨圖片提供｜樂湉設計　文｜陳佳歆

● 客廳以木色濃郁的胡桃木與大理石紋地面形塑出調性，搭配復古設計的燈具與單椅，傳遞出質樸溫潤的氛圍。

Before　　　　　　　　　　　After

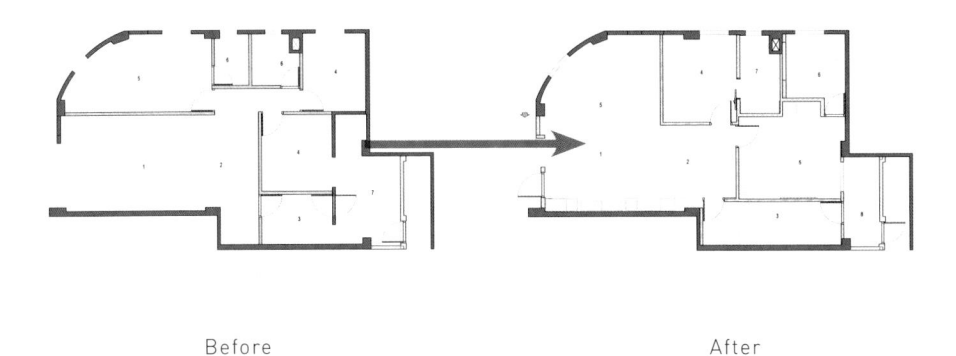

　　從事音樂工作的屋主，將過往出租的老屋收回重新改造，方便照顧年邁行動不便的母親。

　　位在一樓的老屋先前作為辦公室使用，格局並不適合目前的居住需求，原本採光條件就不好，再加上開窗較小，空間感覺很昏暗，天花板也有大樑壓低空間，整體空間需要以未來的使用需求重新調整規劃。

　　考量到長輩生活起居使用的方便性，空間格局設定為雙套房，將原本三房格局移除一間鄰近客廳的主臥，使得客廳空間變得更為寬敞舒適，也因為少了臥房牆面的阻擋，公領域採光面增加有更多光線流入。

　　由於要將兩間次臥改成雙主臥套房，便合理的縮小工作陽台比例放大臥房空間，原本位在角落的臥房則規劃為一間四件式衛浴，幸運的是，因為位在一樓可以從地下樓層改管，免除了一般移動衛浴需要墊高地坪的狀況，少了高低落差的地坪進出步伐更為安全。

　　線條簡單的空間以濃郁的胡桃木作為整面櫃牆，後面隱藏著電視及各種收納，地坪搭配大理石紋地磚，優雅的呈現屬於當代的復古的氛圍。

減少一房打開公領域尺度

依照屋主需求將原本三房格局重新規劃為雙主臥套房，因此移除鄰近客廳的主臥，不但放大公領域也增加採光面，使得空間不再感覺昏暗。

更改次臥為無障礙設計衛浴

原先位在邊間的次臥格局，重新規劃為四件式的主臥衛浴，同時也置入無障礙設計，在浴櫃增加扶手以防滑倒，提升沐浴時的安全性。

選搭磁磚素材營造復古廚房

推開餐廳牆面的隱藏門,看到延續公領域調性的廚房,壁面選擇清爽花磚搭配不銹鋼流理台,與地面紅色陶板磚呼應出當代的復古調性。

CASE 16

HOME DATA

屋齡｜40 年　屋型｜公寓　坪數｜40 坪　家庭成員｜1 人
建材｜鐵件、香杉實木、玻璃磚、木作格柵、榻榻米、特殊漆砂石塗料、木皮、系統櫃

洄遊於前庭後院的簡約日宅

空間設計暨圖片提供｜日作空間設計　文｜Fran Cheng

●將原本房間隔牆敲除，改以黑框玻璃落地窗來打造無隔閡的現代客廳，不僅改善原本陰暗感，庭院白色抿石子圍牆與綠蔭可以遮掩路過目光，也可放下遮光簾保護隱私；而黑磚步道後方的玄關木櫃則有如沙發背牆成為端景。

Before　　　　　　　After

　　身爲上班族的屋主，因期待退休後能有愜意養生的綠意宅院，決定將四十年老屋重新整修，把原本兩房兩廳變得更寬敞，不僅方便年紀更大時在家能自由行動，年節時家人團聚也更舒適。

　　屋主喜歡簡約新日式風格，又想保留舊家具，所以規劃重點之一就是將新舊的日台家具結合，並由此延伸出深色木質調空間。

　　爲了一掃老屋陰暗，先將原本鄰接庭院、擋住光線的房間拆除，沿前庭雙側定位出客廳與和室，並由中間入門處鋪設黑磚走道。

　　接著在走道與和室間立起櫥櫃來區隔與中介客廳，櫥櫃既提供玄關收納、穿鞋與遮掩等需求，也形成回字動線，展現多元生活場景。

　　至於和室外牆改採摺疊拉門，搭配日式緣廊拉近內外距離，親友來訪可在此聊天茶敘，必要時搭配障子門、遮光簾也能做客房用。

　　客廳外牆採黑框全玻璃設計，讓庭院綠景直入室內，搭配包覆舊家具的木牆櫃，打造沉穩且寧靜的簡約廳堂。

白牆木階重建綠植庭院

從庭院外牆開始重建抿石子白牆、灰色鐵門與雨遮，搭配高低階碎石、塑木平台與綠植規劃，滿庭綠意成為戶外、室內都能享受的雅緻庭院。

多功能玄關木櫃與黑岩磚道

由大門向室內延伸一道黑岩磚道，除了可區隔客廳與和室，設計時也考慮未來輪椅進出的寬度及耐髒性、襯托木質感等細節。而左側玄關木櫃既是客廳端景、玄關收納櫃，兩側平台也是穿拖鞋椅區。

儲藏室也是餐酒櫃與端景

走道前方白牆後方為儲藏室，除可收納大型過季物品，也區隔餐廳與主臥兩區。正面搭配五斗櫃與掛畫可作為玄關端景，右側面對餐廳則配置有餐櫃與紅酒櫃。此外，左側客廳木櫥櫃則是請木作包覆舊家具重製設計的裝飾牆。

多功能和室坐擁日式庭院

為了不浪費前庭格局，在落地窗旁規劃了榻榻米和室，搭配摺疊玻璃門與竹林營造日式緣廊情境，也自成回字動線。當親友來訪時不僅可在此茶敘、聊天，左側木櫥櫃拉下就是床鋪，搭配障子門與遮光簾就是間客房。

CASE 17

HOME DATA

屋齡｜約32年　屋型｜電梯大廈　坪數｜10坪　家庭成員｜夫妻
建材｜壓花玻璃、不鏽鋼噴烤漆板、超耐磨地板、實木貼皮

10坪老屋全開放，
重塑光感美學宅

空間設計暨圖片提供｜隹設計　文｜EVA

◉入口廚房向內挪移，與客廳整合，還原開闊的玄關廊道。而原有和室拆除，釋放公領域空間，賦予開闊明亮的視野。

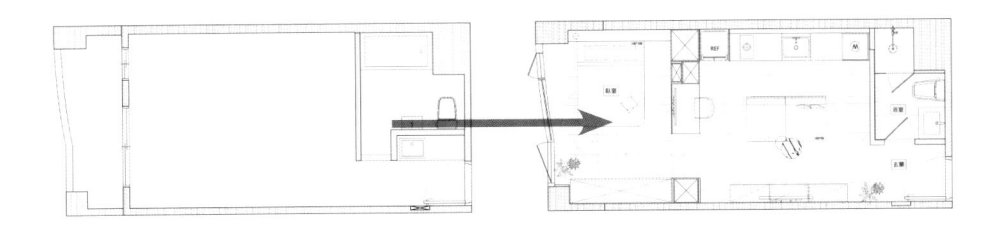

Before　　　　　　　　　　　　After

　　本案為一對年輕夫妻的新居，屋齡三十二年、僅10坪的空間對於採光與動線是一大挑戰。

　　原始廚房位於大門入口，空間侷促影響動線，中央設置和室，光線無法流通，顯得陰暗壓迫，因此整體以改善採光與動線為核心，重塑開放敞亮的格局。

　　首先將廚房向內挪移，釋放玄關領域，維持入口的通暢開闊。中央拆除和室，衛浴縮小讓出空間，客廳與廚房整合後，公領域更通透顯大。而一字型的廚房設計保有順暢的料理動線，也暗藏洗衣設備，完善家務機能。

　　臥室則沿窗設置，特意不做隔間，改以櫃體和書桌區隔，提升空間坪效。櫃體不做置頂，保留上方空間，讓光線得以穿透，同時架高地板隱性圍塑場域，也多了收納空間。

　　收納規劃更講究簡約，運用隱藏式設計與牆面、地面融為一體，維持簡潔俐落的視覺感受。為了讓整體更通透寬敞，臥室與衛浴則選用壓花玻璃，透光不透視的設計讓光線能恣意流動，也能確保隱私。

重塑料理、洗衣家務動線

全室以白色作為主色,有效放大空間,形塑簡潔清新調性。廚房牆面特地選用不鏽鋼烤漆板,不僅便於清潔,也具磁吸功能,能靈活安排收納。下方則藏入洗衣設備,料理、家務一氣呵成。

櫃體不做滿,維持開放通透

臥室沿窗設置,架高地面並以櫃體和書桌作為隔間,不僅工作、收納機能一應俱全,也提升空間使用效率。為了不讓空間過於壓迫,櫃體高度特意不做滿,輔以壓花玻璃區隔,保有穿透視野,整體開闊流暢。

玻璃隔間與門片，保有通透採光

縮小衛浴，拆除浴缸改為淋浴間，將空間釋放給公領域。採用半高隔牆與壓花玻璃區隔，搭配透明門片，有效引入光源，改善陰暗困擾。牆側更貼心設置壁龕開口，能作為夜間微光源使用，不僅行走更安全，也營造溫暖氛圍。

CASE 18

HOME DATA

屋齡｜約 40 年　屋型｜電梯大廈　坪數｜37 坪　家庭成員｜夫妻、2 小孩
建材｜木地板、美耐板、特殊塗料、花磚、灰鏡

老屋格局重整變開闊，
收納加倍升級

空間設計暨圖片提供｜日居設計　文｜Celine

●從餐廚區望向客廳，空間採開放式設計，適當地於天花板、櫃體轉角處融入弧線設計，既有畫龍點睛的視覺效果，也讓整體氛圍更為柔和。

Before　　　　　　　　　　After

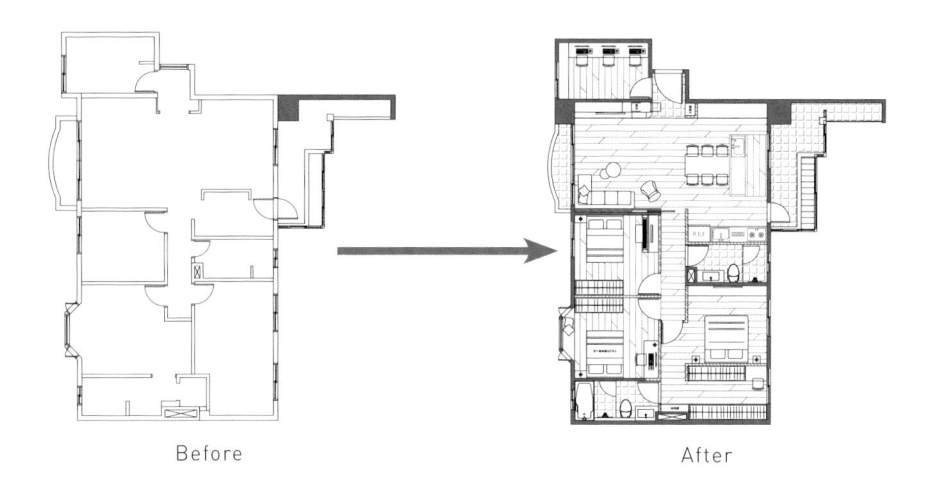

　　這間老屋既有屋況條件還算不錯，可惜的是原始格局存在許多畸零空間，削弱了空間的便利性與通透感。設計師針對這些問題進行全面優化，以滿足兩大兩小的家庭需求，並重新賦予空間生命力。

　　首先是將廚房隔間拆除，讓公領域擁有開闊的尺度，三間臥室也重新經過調整，化解零碎無用的角落，而為了區隔公私領域，特別設置一道電動拉門，門板採用美耐板斜貼設計，既實用又增添視覺層次。

　　原本最令屋主困擾的鞋物收納問題，設計師更利用電視牆側面整合橫拉式旋轉鞋櫃，大幅提升收納機能。

　　客廳天花板的弧線設計則修飾大樑，結合內藏間接燈光，為灰階基調增添柔和氛圍，同樣的弧線元素也運用於玄關轉角，搭配灰鏡與花磚拼貼地面，營造豐富的視覺效果。一方面，開放式餐廚增設中島設計，滿足屋主期望的雙水槽功能，餐邊櫃更貼心預留掃地機器人的收納空間，展現細膩考量。

　　整體設計從格局調整到細節處理，完美融合機能與美感，讓這間老屋以全新面貌展現獨特魅力。

複合式電視牆隱藏大容量鞋櫃

簡單俐落的灰階色電視牆，在於右側轉角處隱藏了鞋櫃，巧妙化解既有格局難以配置鞋櫃的問題，橫拉式設計，內部以旋轉架提高收納量，兩側皆可使用，融合美觀與實用，也最大化玄關收納空間。

弧線天花，柔化空間

弧線天花板內藏間接燈光，巧妙修飾大樑，灰階牆面搭配美耐板斜貼設計電動拉門，與淺色家具共同營造柔和且分區明確的客廳氛圍。

中島餐廚提升機能與開闊性

開放式餐廚增設中島，創造屋主希冀的雙水槽功能，灰階色調與大地色餐邊櫃相得益彰，櫃體貼心預留掃地機器人收納空間，展現機能與美感兼具的設計巧思。

CASE (19)

HOME DATA

屋齡｜約30年　屋型｜公寓　坪數｜20坪　家庭成員｜夫妻
建材｜超耐磨木地板、美耐板、鏡面、長虹玻璃

30年老公寓蛻變溫暖北歐風

空間設計暨圖片提供｜寬月設計　文｜Celine

◉私領域的房門入口採用木作隱藏門設計手法，讓立面更形簡潔俐落，也成為進門後的一道端景。

Before　　　　　　　　After

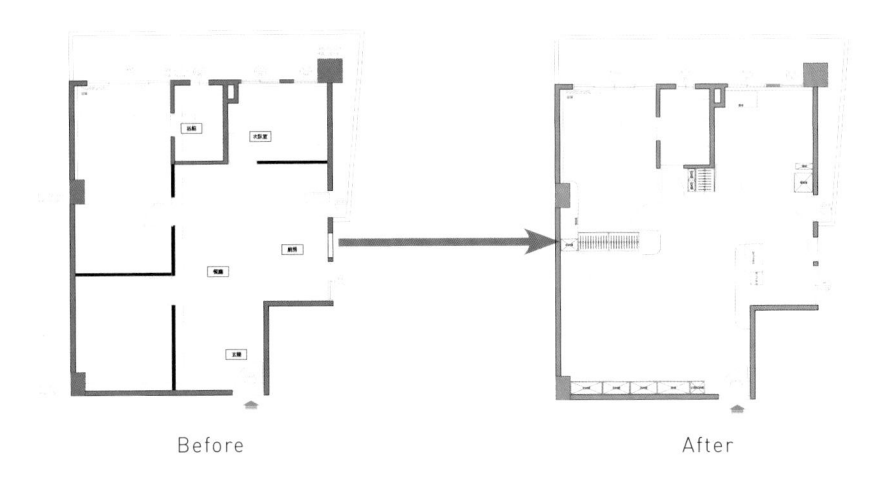

　　這是一間屋齡約三十年的頂樓公寓，原本作為出租用途，因此並沒有客廳，而是隔成一間間臥室，如今夫妻倆決定重新翻修享受退休生活，經過設計師的改造，以「開放尺度」、「光線引入」與「溫暖自然」為核心，徹底改變老屋的陳舊面貌，蛻變為兼具功能性的北歐風現代居所。

　　首先，打破原本封閉的格局，將客廳與廚房連結成為開放式型態，創造出視覺與動線上的通透感，此外也特別讓電視牆與大門同一側，讓人一進門就能感受到寬敞開闊的生活尺度。

　　有別於一般餐廳的配置，改以中島餐檯取代，兼具用餐與書桌功能，並再度提升空間的寬闊性。

　　而中島桌面則選用石紋美耐板，不僅易於清潔，也展現北歐風的簡約質感，另外，長虹玻璃成為亮點，既可引入自然光線，更防止油煙逸散，營造溫暖而通透的氛圍，中島桌面與長虹玻璃更透過弧形細節處理，有效降低碰撞風險，體現人性化考量。

　　材質與色彩上，選用耐磨木地板取代原有地磚，搭配木紋美耐板與白色系統櫃，營造清新溫馨的生活氛圍；鞋櫃與收納櫃採用白色門片搭配鋁框玻璃門，兼具功能性與設計美感，藉由北歐風的簡約與溫暖，給予屋主更加舒適與實用的生活體驗。

開放餐廚串聯客廳打造開闊通透感

將電視牆規劃於入口同一側，加上開放中島餐廚的設計，讓公領域擁有開闊通透的空間尺度，同時滿足屋主在客廳運動的需求。電視立面以簡約白色搭配鋁框玻璃門，兼具實用與美感。

自然溫暖的簡約北歐風

開放式公領域讓光線自然流通，大幅提升舒適性，簡約的北歐風設計，灰色沙發搭配淺木地板，自然溫暖，同時利用沙發轉折過道的轉角處，以開放層架結合半腰櫃體，增加實用性

多功能中島滿足用餐與工作

中島餐廚桌面採用石紋美耐板，耐用且易於清潔，中島兼具用餐與辦公功能，取代傳統餐桌，提升空間靈活性。搭配弧形長虹玻璃，引入光線並防止油煙擴散，讓空間既實用又充滿北歐風的溫馨氛圍。

點綴木紋營造溫暖簡約氛圍

主臥室延續北歐風格，系統床頭板融入溫潤的木作質感，搭配超耐磨木地板營造舒適氛圍。內嵌書桌與開放層架設計提供機能性，提供梳妝或日常閱讀適，加上淺色系牆面與柔和光線，打造出一處溫暖且放鬆的休息空間。

CASE (20)

HOME DATA

屋齡｜約 50 年　屋型｜公寓　坪數｜31 坪　家庭成員｜父母、2 小孩
建材｜超耐磨木地板、樺木、塗料、水磨石磁磚

四房變五房，
打造三代共享的無障礙退休宅

空間設計暨圖片提供｜福研設計　文｜EVA

◉拆除封閉廚房，改為彈性開放的設計，串聯客廳與餐廚，拉伸公領域視野。中央則安排私領域，拓寬長廊以利輪椅通行，空間無障礙。廊道搭配圓拱天花與均勻光源，向上延展空間，不顯壓迫。而兩側則輔以展示收納，避免走道單調狹窄。

Before

After

　　本案爲一間 31 坪老屋，屋主期待讓父母與兄弟有各自的臥室，並增設書房滿足在家工作的需求。同時考量父母退休生活，整體需納入無障礙設計，確保未來居住的舒適與安全。

　　首先拆除封閉廚房，運用中島、拉窗彈性區隔，能依需求改爲開放或獨立廚房，不僅放大客廳、餐廚的公領域，也兼顧社交互動與油煙控制。

　　餐廳則規劃伸縮餐桌，平常能四人使用，親友聚會時可拉長至八人長桌，靈活應對各種情境。

　　爲了確保一家四口的臥寢空間，保留原有的四房格局，拆除主衛，改爲獨立書房，提供安靜舒適的工作環境。

　　而主臥退縮牆面，讓出空間給兩間客衛，其中一間則擴大爲無障礙設計，方便未來輪椅進出。

　　至於中央長廊適度加寬，確保輪椅順暢通行，同時沿牆設置收納與展示空間，不僅提升機能，豐富的視覺層次也讓空間不顯壓迫。而廊道輔以拱型天花向上拉伸視野，搭配光線漫射，有效打破長廊的封閉感。

玄關增設拉門，提升隱私

由於原始玄關與廊道相對，隱私性不足，因此運用洞洞板拉門巧妙區隔，有效避免入門直視走廊，也擴增玄關收納。同時牆面特意留出觀景窗，搭配感應式照明，能作為進出的提示指引。

搭配拉窗，廚房開闊顯大

原有廚房過於封閉狹窄，因此拆除隔間，擴大廚房領域，中島上方設置拉窗、一側搭配門片，有需要時能全然遮擋油煙，半開放的彈性設計讓空間調度更靈活。

伸縮餐桌，使用更有彈性

考量到常有親友聚會的情況，中島暗藏伸縮餐桌，平常的四人餐桌能延伸拉長，容納八人使用，彈性利用的設計，不論人再多都能滿足需求。

次臥設置和室，擴大機能

次臥架高地板增設和室，擴大使用彈性，當親友來訪能轉作客房，空間靈活運用。全室地板則暗藏收納，不僅擴增多元機能，也呈現俐落簡潔的視覺。

CASE (21)

HOME DATA

屋齡│約 37 年　屋型│透天　坪數│65 坪　家庭成員│夫妻、2 小孩
建材│超耐磨地板、實木貼皮、木絲水泥板、系統櫃、石膏磚、玻璃

30 年陰暗老屋，
變身開闊淨白大宅

空間設計暨圖片提供│隹設計　文│EVA　攝影：Studio Millspace

◉ 拆除老舊圍牆，一樓空間退縮，保留雙車位的空間。外牆則重新塗刷，淨白色調形塑清新質感。一側則以石膏磚鋪陳鏤空立面，輔以玻璃維持透亮視野，有效引入光線，也遮擋四周視野確保隱私。

| 1F | 2F | 3F | 1F | 2F | 3F |

Before　　　　　　　　　　　　　After

　　這棟三十多年透天老屋，有著傳統長型街屋的問題，僅在前側有採光，空間陰暗潮濕、通風不佳，再加上停車空間不足。屋主一家四口期待能重新改善老屋體質，打造明亮舒適的居住空間。

　　整體率先從建築外觀著手，拆除老舊圍牆與窗戶，利用玻璃與石膏磚鋪排鏤空外牆，有效納入光線，也能遮擋鄰居視線，增加生活的私密性，而潔白的建築立面也讓視覺更為簡潔俐落。

　　考慮到老屋位處巷弄，停車不易，因此一樓向內退縮，縮減客廳，釋放雙車位的停車空間。同時隔牆則改為落地大窗，後側廚房也同時縮減，打造開放天井，讓光線與空氣能夠深入室內，客廳與餐廚維持通透明亮的視野。

　　而樓梯下方則嵌入儲藏室，並沿樓梯側面安排櫃體，巧妙化解入門直視樓梯的困擾，也擴大收納機能。二樓則規劃開放的起居空間作為家庭的社交中心，形塑工作、運動、閱讀和遊戲共用的多功能場域。主臥則與工作陽臺相連，打造順暢的洗衣、晾衣動線。

樓梯包覆櫃體，活用畸零空間

玄關沿著樓梯安排儲藏室與櫃體，善用樓梯下的畸零空間，也化解樓梯的厚重量體。櫃體刻意不做置頂，保留通透的呼吸感。同時對側沿窗安排矮櫃，也能充當穿鞋椅，滿足多元機能。

開放天井，空間通透流暢

由於長型街屋有著採光不足、通風不良的問題，因此廚房、衛浴向內退縮，打造開放天井，即便位處室內，也能擁有開闊通透的空間感。

開放起居空間
形塑家庭重心

二樓為家庭主要的活動空間，安排起居室彈性運用，能成為大人工作、運動的空間，也是小孩遊戲的場域，形塑親子共用的社交重心。一側則輔以櫃體增添收納，確保完善機能。

主臥強化採光與家務動線

二樓安排主臥，並增設更衣空間，以玻璃拉門區隔確保通透與隱私。而主臥與工作陽臺相連，改善洗衣、晾衣的家務動線。同時拆除舊有雨遮，輔以大面窗戶強化自然採光，打造舒適開闊的臥寢環境。

DESIGNER DATA

十一日晴空間設計
TheNovDesign@gmail.com
116 臺北市文山區木新路三段 243 巷 4 弄 10 號 2F
701 臺南市東區東和路 146 號 3F

日作空間設計
02-2766-6101
rezowork@gmail.com
110 臺北市信義區松隆路 9 巷 30 弄 15 號

日居設計
02-2883-3570
CNdesign250@gmail.com
111 臺北市士林區大東路 162 號 5F

季沃設計
02-2366-0200
info@zwork.com.tw
110 臺北市中正區金門街 34 巷 20 號 1F

隹設計
0919-972-359
dszhui@gmail.com
105 臺北市松山區民生東路五段 99 號 2F

森叄設計
02-2325-2019
service@sngsandesign.com.tw
106 臺北市大安區建國南路二段 171 號 2F

賀澤設計

03-6681-222

hozo.design@gmail.com

302 新竹縣竹北市自強五路 37 號

福研設計

02-2703-0303

happystudio@happystudio.com.tw

106 臺北市大安區安和路二段 63 號 4F

寬月設計

080-055-5848

kuanmoon0800@gmail.com

110 臺北市信義區虎林街 108 巷 138 號

樂湉設計

0975-695-913

Lsdesign16@gmail.com

105 臺北市松山區敦化南路一段 100 巷 26 號 1F

老屋裝修基礎課

2025 年 03 月 01 日初版第一刷發行

編　　著	東販編輯部
編　　輯	王玉瑤
採訪編輯	Celine・Fran Cheng・EVA・喃喃・陳佳歆
封面・版型設計	謝小捲
特約美編	梁淑娟
發 行 人	若森稔雄
發 行 所	台灣東販股份有限公司
	＜地址＞台北市南京東路 4 段 130 號 2F-1
	＜電話＞(02)2577-8878
	＜傳真＞(02)2577-8896
	＜網址＞http://www.tohan.com.tw
郵撥帳號	1405049-4
法律顧問	蕭雄淋律師
總 經 銷	聯合發行股份有限公司
	＜電話＞(02)2917-8022

著作權所有，禁止翻印轉載
Printed in Taiwan
本書如有缺頁或裝訂錯誤，請寄回更換（海外地區除外）。

TOHAN

老屋裝修基礎課 / 東販編輯部作.
-- 初版 . -- 臺北市：
臺灣東販股份有限公司, 2025.03
176　面；17×23 公分
ISBN 978-626-379-778-9（平裝）

1.CST: 房屋 2.CST: 建築物維修 3.CST: 室內設計

422.9　　　　　　　　　　　　114000654

❼ **make sb's way to...** 某人一路前往 (某地)
We made our way to the west coast from Chicago.
我們從芝加哥一路前往西岸。

❽ **even though...** 即使／儘管……
Even though Nick is retired, he always tries to stay active.
儘管尼克已退休了，但他總是保持活躍。

經典旅遊實用句

How about we start by _____ ? 我們先……怎麼樣？
用途：提出旅遊建議，適合討論旅程順序或行程安排。
我們先穿越香港和中國如何？
How about we start by going through Hong Kong and China?
　　　　　　　　　　 booking the flights　訂機票
　　　　　　　　　　 visiting Hong Kong　去香港

經典旅遊實用句

🎯 直接拿來用的旅遊句型
掌握**必備句型**，讓你在旅途中一用就上手，溝通更有自信。

旅人私房筆記

許多背包客會選擇經濟實惠的旅行方式，例如：

| hostel 青年旅館 | 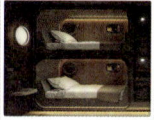 capsule hotel 膠囊旅館 |  workaway 打工換宿 |

| couchsurfing 沙發衝浪（白住在別人家裡） |  house-sitting （屋主不在時）替人看房子 |

📍 文化知識與旅遊小祕訣
提供**文化知識、旅遊小撇步、實用補充**，讓你的旅遊英文更接地氣、不失禮。

Try It Out! 試試看

請依句意在空格內填入適當的字詞

❶ 艾迪在穿越沙漠長途跋涉時遇到響尾蛇。
Eddie encountered a rattlesnake when he was _____ through the desert.

❷ 城市裡的生活相當忙碌。相比之下，生活在鄉下頗輕鬆的。
Life in the city is pretty busy. By contrast, living in the country is quite _____.

Ans ❶ trekking ❷ relaxing

😊 邊看邊寫，馬上練習
複習剛學會的實用字詞，加深記憶不卡關。

旅遊英語輕鬆聊

參考答案請掃描使用說明頁上 QR code 下載

Belle and Alex talk about balancing hard traveling with relaxing activities. What do you think is a good balance between adventure and relaxation when traveling?

💬 簡單開口問答練習
透過簡單口語問答，思考並說出自己的答案，讓學習更貼近**實際使用情境**。
附有參考答案及音檔，好學好用好輕鬆。

AI 功能輔助

三種等級（初學者／中級者／進階者）ChatGPT 指令模組＋對應 SOP，讀者可以依照自己的英文程度選擇最適合的練習方式。
【SOP】如何用 ChatGPT 搭配本書練習英文會話（依程度分級）

準備工具

- 一本《旅遊英語這樣學最有效：65 個情境 × 真人示範 × 互動口說練習》
- 手機（可拍照／截圖／上傳圖片）
- 開啟 ChatGPT（推薦使用 app）

操作步驟（SOP）

Step 1 挑選你想練習的對話內容
翻開書本，找到你想練習的情境對話（如：飯店入住、點餐、搭車等）。

Step 2 拍照或截圖那一頁對話
請確保畫面清晰完整，對話的英文句子都能辨識。

Step 3 打開 ChatGPT 並上傳圖片
在聊天室中上傳剛剛拍好的照片。

Step 4 選擇適合你程度的指令貼上
請根據你的英文程度，複製以下其中一段指令並貼給 ChatGPT ⬇

點此可直接複製文字

- 初學者版指令：簡單對話＋中文解釋

> 你是一位充滿活力且鼓勵學生的英文老師，幫助學生練習英文口說。你會使用我上傳的照片內容作為主要教材，與我練習旅遊英文對話。
>
> 我是初級英文學習者，請用非常簡單的英文和我對話，慢慢來。當我回答錯誤時，請幫我修正，並用中文解釋錯在哪裡。再請你用正確的句子繼續對話。
>
> 請你自然地引導我練習，不要一次給太多句子，也不要讓我有壓力。現在請根據照片中的對話開始角色扮演吧！

● **中級者版指令**：全英文練習 + 引導說更多

> 你是一位熱情且有耐心的英文老師，現在要幫助我進行英文口說練習。請根據我上傳的照片（書中的實境對話）模擬對話情境，跟我一來一往對話。
>
> 我是中級英文學習者，希望用英文自然地表達。如果我犯錯，請用簡單英文幫我糾正，並鼓勵我繼續說下去。請適當引導我延伸回答，例如問我一兩個追問的問題。
>
> 請開始英文對話練習，角色扮演開始！

● **進階者版指令**：挑戰延伸應用 + 強化自然口說

> 你是一位專業的英文老師，請協助我進行進階旅遊情境英文對話練習。請參考我上傳的照片（書中的實境對話）設計一段有挑戰性的角色扮演練習。
>
> 我希望你用自然、生活化的英文和我進行一來一往的會話，並加入一些延伸情境或意外情況（例如：飯店沒房間、航班延誤等），考驗我臨場反應。
>
> 如果我使用不自然或有錯誤的句子，請指出並用英文改寫為更道地的說法。請讓練習更貼近真實旅遊情境，幫我提升實戰能力。

Step 5 開始跟 ChatGPT 對話吧！
你會發現 ChatGPT 會根據你提供的照片與指令，用英文帶你對話、糾正錯誤、甚至鼓勵你多說一點。

🔍 **小建議**

- 初學者可搭配旅遊字詞補給包的單字，幫助聽懂回應內容。
- 中級者可以請 ChatGPT「用更自然的說法幫我改寫這句話」。
- 進階者可要求「請你模擬一個更困難的場景，例如航班取消」。

目錄 Contents

Chapter 01　Plan & Preparation　行前規劃準備 ……… 1

- **Unit 01**　Planning a Trip　計劃旅行 ……………………… 2
- **Unit 02**　Check Your Passport!　護照到期怎麼辦？……… 6
- **Unit 03**　Applying for a Visa　申辦簽證 ………………… 10
- **Unit 04**　Getting Advice on a Vacation　旅行社的推薦行程 …… 14
- **Unit 05**　Flight Ticket Reservations　來去訂機票吧 …… 18
- **Unit 06**　Choosing the Best Flight　選擇最佳航班 ……… 22
- **Unit 07**　Hotel Reservations　預訂飯店 ………………… 26
- **Unit 08**　Packing for a Trip　打包行李 …………………… 30

Chapter 02　Taking a Flight　搭機 ……………………… 35

- **Unit 09**　Boarding and Check-in　報到登機 ……………… 36
- **Unit 10**　Where's My Passport?!　我的護照呢？！……… 40
- **Unit 11**　Going through Security at the Airport　通過機場安檢 …… 44
- **Unit 12**　Getting Bumped to the Next Flight　候補機位 …… 48
- **Unit 13**　Flight Delay　班機延誤怎麼辦 ………………… 52
- **Unit 14**　Ugh! Flight Canceled!　航班取消真不便！…… 56
- **Unit 15**　Missing One's Flight　「機」不可失 …………… 60

Unit 16	Finding Your Seat on the Plane	找座位	64
Unit 17	A Poorly-Behaved Passenger	飛機上的奧客	68
Unit 18	In-Flight Service	先生，您還需要什麼呢？	72
Unit 19	Asking a Flight Attendant for Help	向空姐求助	76
Unit 20	Last-Minute Gifts	最後才想到買禮物	80
Unit 21	Getting Directions through the Airport	機場裡找路	84
Unit 22	Customs Declaration	海關申報	88
Unit 23	Missing Luggage	消失的行李	92

Chapter 03　Accommodations　住宿　97

Unit 24	Checking In at a Hotel	飯店入住登記	98
Unit 25	Reservation Trouble	入住搞烏龍	102
Unit 26	Room Service	客房服務	106
Unit 27	Complaining to Hotel Staff	天有不「廁」風雲！	110
Unit 28	Cleaning the Room	打掃客房	114
Unit 29	Checking Out of a Hotel	從飯店退房	118
Unit 30	Taking the Hotel Shuttle	搭乘飯店接駁車	122

Chapter 04　Transportation　交通　127

Unit 31	When in Rome	羅馬輕鬆遊	128
Unit 32	Renting a Car	租輛車子好代步	132
Unit 33	Filling Up the Tank	加滿油	136

Unit 34	Buying a Train Ticket	購買火車票	140
Unit 35	Taking a Taxi	搭計程車	144
Unit 36	Taking the Bus	搭公車	148
Unit 37	Getting on the Wrong Bus	公車迷途記	152

Chapter 05　Sightseeing　觀光　157

Unit 38	Touring in the UK	英國旅遊趣（詢問旅客資訊站）	158
Unit 39	Buying Tickets	買票	162
Unit 40	Entering an Amusement Park	驗票入遊樂園	166
Unit 41	Fun at the Amusement Park	遊樂園小約會	170
Unit 42	Touring a Museum	參觀博物館	174
Unit 43	Taking in a Theater Show	一起去看戲	178
Unit 44	Preparing for a Skiing Trip	準備滑雪趣	182
Unit 45	Taking Pictures	拍照	186

Chapter 06　Food　美食　191

Unit 46	Enjoying Local Food	品嚐當地美食	192
Unit 47	Having Afternoon Tea	下午茶	196
Unit 48	Ordering Food at a Diner	在餐廳點餐	200
Unit 49	Trying New Foods	嘗試新食物	204
Unit 50	Tipping Culture	小費文化知多少	208
Unit 51	Paying the Check	結帳	212

Chapter 07　Shopping　購物 …… 217

- Unit 52　Exchanging Currency　兌換外幣 …… 218
- Unit 53　Trying on Clothes　試穿衣服 …… 222
- Unit 54　Bargaining in a Store　殺價 …… 226
- Unit 55　Requesting a Return or Refund　要求退貨退款 …… 230
- Unit 56　Duty-Free Shopping at the Airport　免稅店購物樂 …… 234
- Unit 57　Getting a Tax Refund　申請退稅 …… 238

Chapter 08　Emergency　緊急狀況 …… 243

- Unit 58　Making an Appointment　掛號 …… 244
- Unit 59　Talking with the Doctor　向醫生說明病情 …… 248
- Unit 60　A Quick Stop at the Pharmacy　去藥局領藥 …… 252
- Unit 61　Reporting a Crime　警察局報案 …… 256
- Unit 62　Checking the Lost and Found　失物招領處 …… 260
- Unit 63　Asking for Directions　問路 …… 264
- Unit 64　Looking for a Restroom　找廁所 …… 268
- Unit 65　Losing Your Passport　遺失護照 …… 272

Chapter 01

Plan & Preparation
行前規劃準備

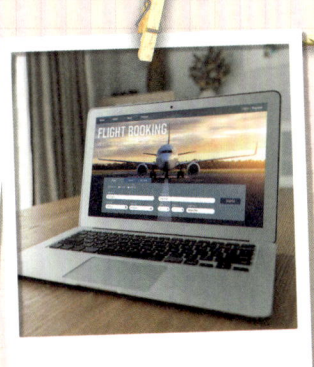

Unit 01
Planning a Trip
計劃旅行

Unit 02
Check Your Passport!
護照到期怎麼辦？

Unit 03
Applying for a Visa
申辦簽證

Unit 04
Getting Advice on a Vacation 旅行社的推薦行程

Unit 05
Flight Ticket Reservations
來去訂機票吧

Unit 06
Choosing the Best Flight
選擇最佳航班

Unit 07
Hotel Reservations
預訂飯店

Unit 08
Packing for a Trip
打包行李

1

Unit 01 Planning a Trip
計劃旅行

實境對話 GO！

B Belle（貝兒）　　**A** Alex（艾力克斯）

Alex and Belle are planning an around-the-world trip.

B This is going to be **terrific**. We're going to go all across **the globe**.

A I can't wait. Did you tell your parents yet?

B They said once we know where we are going and for how long, we can talk about it **in depth**.

A My parents don't believe we will be able to do it, so I'**m out to** prove them wrong. Where do you want to start?

B Since we are in Asia, I was thinking we can spend our first three months **trekking** around Southeast Asia.

A Perfect. I read that we should balance our hard traveling with stays at **relaxing** beaches.

B Sounds like a dream come true. How about we start by going through Hong Kong and China; then we can **make our way to** Thailand and Vietnam.

A After that, should we go south to Indonesia and Australia or west to Burma, Nepal, and India?

2

B That's a good question. **Even though** we are traveling around the world, we will have to pick and choose based on time and budget. There are going to be a lot of places we won't be able to visit.

艾力克斯和貝兒在規劃他們的環遊世界旅行。

B 這趟旅程將會很讚。我們就要環遊世界了。

A 我等不及了。妳跟妳父母親說了嗎？

B 他們說一旦我們確定要去哪裡和去多久後，我們就可以來仔細深入討論一下。

A 我父母不相信我們能環遊世界，所以我一定要證明他們錯了。妳想從哪個國家開始？

B 既然我們人在亞洲，我想我們可以花前三個月的時間在東南亞徒步旅行。

A 太棒了。我在書上看到說我們應該要在令人放鬆的沙灘上待幾天來均衡一下艱辛的旅程。

B 這聽起來就像美夢成真。我們先穿越香港和中國如何，然後我們可以走到泰國和越南。

A 之後，我們是否該往南去印尼及澳洲，或是往西去緬甸、尼泊爾和印度？

B 這是個好問題。雖然我們是要環遊世界，但我們還是得依時間和預算來挑選地點。會有很多地方是我們無法去的。

旅遊字詞補給包

❶ **terrific** [təˋrɪfɪk] *a.* 極好的，極棒的
This five-star restaurant is famous for its terrific steaks.
這家五星級餐廳以好吃的牛排聞名。

❷ **the globe** [glob] *n.* 地球，世界

commercial [kəˋmɝʃəl] *n.* 商業廣告

Through this **commercial**, we can reach people around the globe.
透過這支廣告，我們就能讓全世界的人認識我們。

❸ **in depth**　深入地；詳細地
The detective didn't investigate the petty theft in depth.
該警探並沒有深入調查這起小竊盜案。

petty theft [ˋpɛtɪ ˋθɛft] *n.* 輕微竊盜

CH 01

Plan & Preparation・行前規劃準備

3

❹ **be out to V**　　試圖 / 力圖（做）……
The politicians are out to change the education system in their country.
政治人物們試圖改變他們國家的教育體系。

❺ **trek** [trɛk] *vi.* 長途跋涉，翻山越嶺　⟶　動詞變化為：trek, trekked, trekked, trekking
It took the mountain-climbing group two days to trek across the rainforest.
這個登山團隊花了兩天跋涉橫越這片雨林。

❻ **relaxing** [rɪˋlæksɪŋ] *a.* 令人放鬆的，輕鬆的
The hot spring was quite relaxing and washed away Sam's troubles.
這溫泉令人相當放鬆，也把山姆的煩惱洗滌一空。

❼ **make sb's way to...**　　某人一路前往（某地）
We made our way to the west coast from Chicago.
我們從芝加哥一路前往西岸。

❽ **even though...**　　即使 / 儘管……
Even though Nick is retired, he always tries to stay active.
儘管尼克已退休了，但他總是保持活躍。

經典旅遊實用句

How about we start by _____?　　我們先……怎麼樣？

用途：提出旅遊建議，適合討論旅程順序或行程安排。

我們先穿越香港和中國如何？
How about we start by going through Hong Kong and China?
　　　　　　　　　　　　booking the flights　　訂機票
　　　　　　　　　　　　visiting Hong Kong　　去香港

旅人私房筆記　　許多背包客會選擇經濟實惠的旅行方式，例如：

hostel
青年旅館

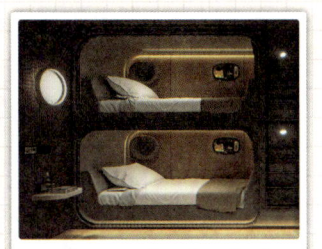

capsule hotel
膠囊旅館

workaway
打工換宿

couchsurfing
沙發衝浪（白住在別人家裡）

house-sitting
（屋主不在時）替人看房子

Try It Out! 試試看　　請依句意在空格內填入適當的字詞

❶ 艾迪在穿越沙漠長途跋涉時遇到響尾蛇。
Eddie encountered a rattlesnake when he was _____ through the desert.

❷ 城市裡的生活相當忙碌。相比之下，生活在鄉下頗輕鬆的。
Life in the city is pretty busy. By contrast, living in the country is quite _____.

　　　　　　　　　　　　　　Ans　❶ trekking　❷ relaxing

旅遊英語輕鬆聊　　參考答案請掃描使用說明頁上 QR code 下載

Belle and Alex talk about balancing hard traveling with relaxing activities. What do you think is a good balance between adventure and relaxation when traveling?

CH 01　Plan & Preparation・行前規劃準備

5

Unit 02 Check Your Passport!
護照到期怎麼辦？

實境對話 GO！

C Chris（克理斯） **H** Helena（海倫娜）

Chris and Helena are discussing their next trip at Chris' house.

C I think **it's about time to** plan our next trip. Where do you think we should go?

H I was thinking of a **tropical** place. How about Bali?

C Bali sounds amazing. But before we **get** too **carried away** picturing ourselves on the beach, we should check our passports to make sure they're still **valid**.

H Good point. Let me grab mine. Oh, no! It looks like my passport is about to **expire**.

C Yikes, that could be a problem. You'll have to **renew** it before we do anything else. It took a little bit of time to renew mine last time.

H You're right. I'll have to gather some documents and **go through** the **application** process, right?

C Yeah, you need to **submit** the required paperwork and pay the fee. It'll take about 10 workdays, so just make sure to do it as soon as possible.

H I see. I'll get my **renewal** documents sorted as soon as I get home.

C Good thinking. Let me know if you need any help. I'll look at **accommodations in the meantime** and send you what I find.

H Thanks! What would I do without you?

克理斯和海倫娜正在克理斯家討論他們的下一次旅行。

C 我想是時候來計劃我們的下一次旅行了。妳覺得我們應該去哪裡？

H 我想去熱帶的地方。峇里島怎麼樣？

C 峇里島聽起來很棒。但在我們沉浸在身處海灘的想像前，我們應該先檢查一下護照，確保它們都還在有效期內。

H 這倒是。我來看看我的護照。喔，糟了！看來我的護照快要到期了。

C 唉呀，那可能會是個問題。在我們做任何規劃前妳得先更新護照。我上次更新護照時花了一點時間。

H 你說得對。我必須準備一些文件並提交申請對吧？

C 對，妳需要繳交所需資料並支付費用。這大概會花十個工作天，所以要確保盡快申請。

H 我明白了。我一回家就會把要更新的文件準備好。

C 好主意。如果妳需要任何幫忙，隨時告訴我。同時我會看一下住宿並將我找到的結果傳給妳。

H 謝啦！沒有你我該怎麼辦？

旅遊字詞補給包

❶ passport [ˈpæsˌpɔrt] *n.* 護照
Make sure to take your passport when you go to the airport.
前往機場時請務必攜帶護照。

❷ it's about time to V　　該是做……的時候了
It's about time to pick my daughter up from school.
到了該去學校接我女兒的時候了。

❸ tropical [ˈtrɑpɪkḷ] *a.* 熱帶的
The tropical climate in Hawaii attracts many tourists each year.
夏威夷的熱帶氣候每年都吸引許多遊客。

❹ get / be carried away　　興奮得失去控制
Sarah tends to get carried away with excitement when playing games.　玩遊戲時，莎拉往往會興奮得失去控制。

❺ valid [ˈvælɪd] *a.* 有效力的；令人信服的
This discount card is only valid until the end of the month.
這張折扣卡效期只到本月底。

❻ expire [ɪkˈspaɪr] *vi.* 期滿，過期
Don't drink the milk. It expired two days ago!
別喝這瓶牛奶。它兩天前就過期了！

❼ renew [rɪˈnju] *vt.* 更新
renewal [rɪˈnjuəl] *n.* 更新
Lisa needs to renew her library card before she can borrow more books.　麗莎需要更新她的借書證才能借更多書。
Peter wants to discuss the renewal of his contract with his manager.
彼得想與他的主管討論續約的事。

❽ go through...　　經歷 / 通過……
If you want to get the job, you have to go through the interview process.　如果你想得到這份工作，你必須經歷面試過程。

❾ application [ˌæpləˈkeʃən] *n.* 申請
Luke's application for the job has been successful.
路克的求職申請已經成功通過。

❿ submit [səbˈmɪt] *vt.* 繳交 & *vi.* 屈服，服從　→　動詞變化為：submit, submitted, submitted, submitting
Jenna submitted her essay before the deadline.
珍娜在截止日期前交出短篇論文。

⓫ accommodation [əˌkɑməˈdeʃən] *n.* 住宿（美式英語中常用複數）
Are there any accommodations available near the beach?
海灘附近有住宿嗎？

⓬ in the meantime　　在此同時
meantime [ˈminˌtaɪm] *n.* 期間
You just focus on studying. In the meantime, I'll start preparing dinner.　你只需要專心讀書。同時，我會開始準備晚餐。

經典旅遊實用句

My passport is... 我的護照……。

用途：說明護照快到期了。

我的護照<u>快要到期了</u>。
My passport is about to expire.
　　　　　　 only valid for 2 more months　只剩兩個月到期

旅人私房筆記

換發護照，一步一步不慌張！

Step 1　Check passport validity　檢查護照效期
- 少於 6 個月？→ 開始換發流程！

Step 2　Gather documents　準備文件
- 舊護照　• 護照照片　• 身分證(未滿 14 歲附戶口名簿)

Step 3　Make an appointment or visit in person　預約 or 親辦
- 網路預約 (外交部官網)　• 臨櫃辦理

Step 4　Pay the fee　繳費
- 新臺幣 1,300 元　• 未滿 14 歲者為新臺幣 900 元

Step 5　Wait about 10 business days　等候約 10 個工作天
- 拿到新護照 → 可以放心訂機票啦！

Try It Out! 試試看

請將下列字詞組合成一正確通順的句子

Jane _____ (got / away / when / carried) she won the race.

Ans　got carried away when

旅遊英語輕鬆聊

參考答案請掃描使用說明頁上 QR code 下載

You're excited for your next trip, but then you realize your passport has expired. Do you panic, or turn it into a mini adventure to get it renewed?

Unit 03 Applying for a Visa
申辦簽證

實境對話 GO!

B Brodie（布洛迪）　　**A** Annie（安妮）

> Brodie, a Canadian **citizen** living in Taiwan, is applying for a Chinese tourist visa while on a business trip to Hong Kong. Annie is working at the visa **booth** in the airport.

B Hello. I'm planning on visiting China in about a month and I'd like to apply for a tourist visa.

A Sure. There is a new Chinese tourist visa that **is good for** 10 years.

B Oh, great. I can see myself visiting China a few times within that time.

A OK, then I'll **process** the 10-year tourist visa for you. May I have your passport, please?

B Yes, here it is.

A Thank you. OK, while I make a copy of it, please **fill out** this application **form**.

B Sure. How much will it cost to process the visa?

A For Canadian citizens, the **fee** is $610 Hong Kong dollars. That's about $104 Canadian dollars.

B Ah, OK, great. That's actually less expensive than I expected. When will I **be able to** pick up my visa?

A It'll take four business days to process. If you need it sooner, you can apply for the **express** service and get it within two days, but there's a fee.

B No worries. I'm here all week for work, so I'll just pick it up at the end of my trip.

A If you have any questions, here is our business card with our **contact information**.

B Great. Thanks for your help!

住在臺灣的加拿大公民布洛迪在香港出差期間正在申請中國觀光簽證。安妮在機場的簽證處工作。

B 哈囉。我計劃大約一個月後去一趟中國，我想要申請觀光簽證。

A 好的。新的中國觀光簽證有效期限為十年。

B 喔，好極了。那我會在有效期間內多遊幾次中國。

A 好，那我這就為您申辦十年簽證囉。可以給我您的護照嗎？

B 好的，給妳。

A 感謝。好，我在印護照影本時，請填好這張申請表格。

B 好的。申辦簽證要花多少錢？

A 加拿大公民需付港幣六百一十元。大約加幣一百零四元。

B 啊，那太好了。那其實比我預期的還便宜。我何時可以來拿簽證？

A 申辦需要四個工作天。如果您想快點拿到，可以申請急件處理，兩天內就可拿到，但須加收費用。

B 沒關係。我整週都待在這裡工作，所以出差結束前拿到簽證即可。

A 如果您有任何問題，這是我們的名片和聯絡資訊。

B 好極了。感謝妳的幫忙！

旅遊字詞補給包

❶ citizen [ˈsɪtəzn̩] *n.* 公民，國民，市民
A good citizen always obeys the law.
好國民總是守法。

❷ booth [buθ] *n.* 服務櫃檯；亭子
I stopped by the ticket booth to buy a ticket for the concert.
我路過售票亭時買了張音樂會的票。

❸ be good for + 一段時間　　在……的期間都有效／適用
The ticket is good for three months.
這張票三個月內都有效。

❹ process [ˈprɑsɛs] *vt.* 處理 & *n.* 過程
The computer takes a few seconds to process the data.
電腦需要幾秒鐘來處理這些資料。

❺ fill out / in a form　　填寫表格
Before the job interview, you have to fill out an application form.
工作面試前，你必須先填申請表。

❻ fee [fi] *n.* (手續) 費用
We must charge you a fee of US$10 for the service.
這項服務我們必須向您收手續費十美元。

❼ be able to V　　能夠／會(做)……
Amy is able to handle the problem alone.
艾咪能獨自處理這個問題。

❽ express [ɪkˈsprɛs] *a.* 快速的；快遞的
Sarah sent the package by means of express mail.
莎拉用快遞寄出那個包裹。

❾ contact information　*n.* 聯絡資訊
Please write your contact information on this paper.
請把你的聯絡資料寫在這張紙上。

經典旅遊實用句

I can see myself [動作] within that time.
我會在有效期限內……。

用途：表示對簽證期限的肯定與規劃。

我會在有效期間內多遊幾次中國。
I can see myself <u>visiting China a few times</u> within that time.
<u>visiting again</u>　　再次到訪
<u>traveling there</u>　　去那裡旅遊
<u>going on business trips</u>　　去出差

旅人私房筆記

簽證百百種，你適用哪一種？

Visa Types

Tourist 觀光簽證
- 觀光
- 不可工作

Business 商務簽證
- 商務訪問
- 不可打工

Student 學生簽證
- 語言學校
- 長期居留

Work 工作簽證
- 僱主申請
- 可合法工作

Transit 過境簽證
- 飛行轉機
- 通常幾日內有效

Try It Out! 試試看

請依句意在空格內填入適當的字詞

我的護照十年內都有效，所以我在需要換照之前還有很多時間。
My passport _____ _____ _____ 10 years, so I have plenty of time left before I need to renew it.

Ans is good for

旅遊英語輕鬆聊

參考答案請掃描使用說明頁上 QR code 下載

You planned to stay in Hong Kong for only three days, but the tourist visa takes four business days to process. Would you pay the express fee or just drink Yuenyeung and wait?

Unit 04 Getting Advice on a Vacation
旅行社的推薦行程

單元音檔 013-016

對話影片 Unit 4

實境對話 GO！

H Heath（希斯）　　**T** Tess（黛絲）

Heath is talking with a travel agent, Tess, about planning a vacation.

H My girlfriend and I really want to visit Africa, but we're not sure where to go.

T No problem. You can **count on** me. There are so many tourist attractions for you to choose from!

H Do you have any advice for people going for the first time?

T Well, the most popular tours are to Kenya or Tanzania. Everyone who visits loves the wildlife **safari**. Lots of people also camp on the **savannah**.

H Oh, wow! That sounds **marvelous**. I would love that, but I'm afraid my girlfriend is not into camping.

T Well, there are lots of other choices and beautiful places to see. Have you thought about visiting Egypt?

H Hmm, I've always wanted to **see** the pyramids **for myself**. But, **to be honest**, I'm not a big fan of deserts.

T Well, then I guess Egypt and Morocco are out. Are you hoping to go to any beaches? Maybe try some surfing?

14

H Yeah, we both love beaches. I can also surf a bit.
T Then how about the Ivory Coast or Madagascar?
H Yeah, maybe. We'd also like to do some hiking and see some wild animals, though.
T In that case, Madagascar will be perfect for you! Here, I'll show you some tour packages.

希斯正在與旅行社職員黛絲討論度假行程。

H 我和我女朋友非常想要去非洲，但我們不確定可以去哪裡。
T 沒問題。包在我身上。有非常多的觀光勝地可供你挑選。
H 妳有什麼建議給第一次去的人嗎？
T 這個嘛，最熱門的旅遊地點都是去肯亞或是坦尚尼亞。每個去的人都非常喜愛野生動物的探險之旅。許多人也會在非洲大草原上野營。
H 哇！聽起來非常棒。我會很喜歡，不過我女友恐怕不是這麼喜歡露營。
T 這個嘛，還是有很多美麗的景點跟其他選擇可以看看。你們有想過要去埃及嗎？
H 嗯……我一直都很想要親眼看看金字塔。但老實說，我並不是很喜歡沙漠。
T 那麼我想可以淘汰掉埃及跟摩洛哥。你們會想要去海灘嗎？也許可以試試衝浪？
H 好耶，我們兩個都很喜歡海灘。我自己也會一點衝浪。
T 那象牙海岸或馬達加斯加如何？
H 或許可以。不過我們也想徒步旅行和看一些野生動物。
T 那樣的話，馬達加斯加會非常適合你們！來，我給你看一些套裝行程。

旅遊字詞補給包

❶ **advice** [əd'vaɪs] *n.* 建議，忠告（不可數）
advise [əd'vaɪz] *vt.* 勸告，提供建言
You should seek legal advice before you sign the contract.
在簽合約之前，你應該先尋求法律建議。
The doctor advised my grandfather not to climb too many stairs.
醫生建議我外公不要爬太多樓梯。

❷ **count on...**　　依靠 / 依賴……
Jane is not a reliable person, so you can't count on her to hand in her homework on time.
珍不是一個值得信賴的人，所以你不能指望她按時交功課。

❸ **safari** [səˈfɑrɪ] *n.* (通常在非洲的) 狩獵、探險之旅
During our safari in Africa, we saw lions, elephants, and giraffes up close.　在我們的非洲野生動物之旅中，我們近距離看到了獅子、大象和長頸鹿。

❹ **savannah** [səˈvænə] *n.* 熱帶的稀樹大草原
The savannah is dry and wide, with few trees.
草原乾燥而廣闊，樹木很少。

❺ **marvelous** [ˈmɑrvələs] *a.* 令人讚嘆的，極好的
The sunrise from the mountain was a marvelous sight.
山上的日出是一個令人讚嘆的景象。

❻ **see (...) for oneself**　　親眼去看 / 親自求證 (……)
Amy couldn't believe her friend's cat could do a backflip, so she had to go see for herself.
艾咪不相信她朋友的貓可以後空翻，所以她要去親眼看看。

backflip [ˈbækˌflɪp] *n.* 後空翻

❼ **to be honest**　　老實說，說實話
To be honest with you, I think the chicken soup you cooked tastes awful.　老實跟你說，我覺得你煮的雞湯很難喝。

經典旅遊實用句

To be honest, I'm not a big fan of [環境 / 類型……].
老實說，我不太喜歡……。

用途：委婉表達偏好。

老實說，我不太喜歡沙漠。
To be honest, I'm not a big fan of deserts.
　　　　　　　　　　　　　　cold weather　　寒冷天氣
　　　　　　　　　　　　　　crowded cities　　擁擠的城市

旅人私房筆記

探索非洲！常見的非洲旅遊

desert camping 沙漠露營
- 摩洛哥和埃及的特色活動，遊客可以在沙漠中露營並騎駱駝。

beaches 海灘
- 非洲的象牙海岸和馬達加斯加擁有美麗的海灘，適合衝浪和放鬆。

safari 野生動物之旅
- 肯亞與坦尚尼亞的特色活動，適合喜歡動物的遊客。

CH 01

Plan & Preparation・行前規劃準備

Try It Out! 試試看
請依句意圈出適當的字詞

如果你不聽我的建議，你就會後悔。
If you don't take my (advice / advise), you will regret it.

Ans　advice

旅遊英語輕鬆聊
參考答案請掃描使用說明頁上 QR code 下載

Would you rather go camping in the wild or stay at a fancy hotel? Why?

17

Unit 05 Flight Ticket Reservations
來去訂機票吧

實境對話 GO！

J Jenny（珍妮）　　S Sam（山姆）

Jenny is calling Sam about booking a flight for their planned trip to Australia.

J Hey, so you know that **promotion** you were telling me about?

S The one that Koala Airlines is having this month?

J Yeah, that one. I'm having trouble booking a flight for some reason. Did you buy your ticket yet?

S No, not yet. I am going to do it tonight. Maybe I can just book ours together.

J That would be great!

S OK, just send me your basic travel information.

J What will you need?

S I'll need your full legal name, like it says on your passport. I'll also need your passport number and probably your date of birth as well.

J OK, I'm messaging all that information to you now.

S Got it! All right, I'll check out the flights now. It looks like the cheapest one **on sale** is NT$18,000.

J That's not too bad. All the ones I found were well over NT$20,000.

S Well, there is a catch, though.

J Oh, no. What is it?

S: With this flight, there's a 23-hour **layover**.

J: What?! That's nearly a whole day! What would we even do during that time?

S: I'm sure there's an airport **lounge** or even an airport hotel we could pass the time in.

J: I**'m** not so **keen on** spending that much time in the airport. Wait, which city is this stop in?

S: Kuala Lumpur. I've never been there.

J: Awesome! I have two friends who live there. We can meet up with them for part of the day. Plus, that would be a great chance to **explore** the city.

S: Then this **itinerary** is not so bad after all. Shall we **go for** this flight then?

J: Yeah, let's book it!

珍妮打電話給山姆詢問關於他們計劃前往澳洲旅遊的航班。

J: 嘿，你記得你之前跟我說過的機票促銷嗎？

S: 是可艾拉航空這個月的促銷嗎？

J: 對，就是這個。我在訂機票的時候遇到了一些問題。你買機票了嗎？

S: 沒，還沒呢。我準備今晚來訂。也許我可以幫我倆一起訂。

J: 那就太棒了！

S: 好，就傳妳的個人旅遊基本資料給我就好了。

J: 你需要哪些資料呢？

S: 我需要妳的法定全名，就像妳護照上寫的名字一樣。我還需要妳的護照號碼，或許妳的出生年月日也一起給我好了。

J: 好喔，我正在傳這些資料給你。

S: 收到！好，我現在來查查航班。看起來最便宜的特價機票是新臺幣一萬八千元。

J: 那還不錯。我找到的都超過新臺幣兩萬元。

S: 嗯，但這種機票有個潛在問題。

J: 噢，不會吧。怎麼了？

S: 這航班中途會停留二十三個小時。

J: 什麼？！那幾乎是一整天了！那我們這段時間要做什麼啊？

19

S 我確定一定有機場休息室或機場附設旅館讓我們可以在那裡打發時間。

J 我並沒有很想在機場裡待那麼長的時間耶。等等，會停在哪個城市啊？

S 吉隆坡。我還沒去過那裡。

J 那太棒了！我有兩個朋友就住在那裡耶。我可以利用當天的部分時間跟他們聚聚。而且這也是探索那座城市的好機會。

S 那這個行程其實還算是不錯。那我們要訂這趟航班嗎？

J 耶，就訂啦！

旅遊字詞補給包

❶ reservation [ˌrɛzɚˈveʃən] *n.* 預約，預訂
It's better to make a reservation before going to the restaurant.
去那家受歡迎的餐廳前最好先預約一下。

❷ promotion [prəˈmoʃən] *n.* (產品的) 促銷，推銷
There is always a promotion on the cooked food after 6:00 p.m. in the supermarket. 下午六點後超市總會有熟食促銷活動。

❸ on sale 　　廉價出售；正在銷售
That watch is on sale for 100 dollars. 那只手錶現在特價一百美元。

❹ layover [ˈleˌovɚ] *n.* 長途飛行中於某地的停留
We had a five-hour layover in Tokyo before our next flight.
我們在東京轉機停留了五個小時才搭上下一班飛機。

❺ lounge [laundʒ] *n.* (飯店、機場、劇院等的) 等候室、休息廳
We waited in the lounge and had free snacks.
我們在休息室邊等邊吃免費點心。

❻ be keen on... 　　對……很喜歡，熱衷於……
Andy is keen on fishing, so he goes to the lake every weekend.
安迪熱愛釣魚，所以他每個週末都會到湖邊。

❼ explore [ɪkˈsplɔr] *vt.* 探索；探險
The students explored the museum with the help of the guide.
學生們藉著解說員的幫助探索博物館。

❽ itinerary [aɪˈtɪnəˌrɛrɪ] *n.* 旅行計畫；預定行程
Our itinerary includes hiking, sightseeing, and a boat tour.
我們的行程包括健行、觀光和搭船遊覽。

❾ **go for sth**　　選擇某事物；喜歡某事物
Between the fruit salad and Caesar salad, I think I'll go for the fruit one.　　在水果沙拉與凱撒沙拉中，我想我會選擇水果沙拉。

經典旅遊實用句

I'm having trouble...　　我在……遇到了一些問題。

用途：表達在預訂上遇到困難。

我在訂機票的時候遇到了一些問題。
I'm having trouble booking a flight.
　　　　　　　　　reserving a hotel　　訂飯店
　　　　　　　　　finding a good deal　　找划算的優惠

旅人私房筆記

Layover vs Stopover vs Transit

✈	layover	短暫中途停留	停留通常少於 24 小時	通常可進出機場
✈	stopover	較長停留	停留超過 24 小時（國際）	可進出機場
✈	transit	過境	停留時間很短	純轉機，不入境

Try It Out! 試試看　　請依句意選出適當的字詞

Maggie is keen _____ outdoor activities like camping and fishing.
Ⓐ to　　Ⓑ in　　Ⓒ on　　Ⓓ for

Ans Ⓒ

旅遊英語輕鬆聊　　參考答案請掃描使用說明頁上 QR code 下載

Would you book a flight with a 23-hour layover if the ticket were super cheap?

CH 01　Plan & Preparation・行前規劃準備

21

Unit 06 Choosing the Best Flight
選擇最佳航班

實境對話 GO！

- **N** Naomi（娜歐蜜）
- **P** Pete（彼特）

Naomi is looking up flights on her laptop when Pete walks by.

N Hey, Pete. I'm **having a hard time** deciding which flight to book for my **upcoming** trip. I want to visit my sister in New York City. Can you help me out?

P Hey, Naomi. One of the most important things for me is the flight time. I always try to choose a flight that fits my schedule so that I won't **be in a rush** after arriving.

N That makes sense. But what about price? Don't you think that is important, too?

P Of course! The price is always an important factor, but I also consider the airline's **reputation**. It's important to choose an airline with a good track record for **reliable departures** and arrivals, especially if you have connecting flights.

N Good point. I also like to consider the **baggage allowance** and whether the airline offers any additional services like in-flight entertainment or meals.

22

P Don't forget to check the seating arrangements, too. Some airlines offer larger space or comfortable seating options, which can **make a** big **difference**, especially on longer flights.

N That's a good idea. Thanks for the tips, Pete. I think I can make my decision now.

P I'm glad to help. I hope you have a nice time with your sister.

當彼特經過時，娜歐蜜正在用筆電查找航班。

N 嘿，彼特。我很難決定要訂哪一家航班來安排即將到來的旅行。我想去紐約找我姊。你能幫我一下嗎？

P 嗨，娜歐蜜。對我來說最重要的事情之一就是航班時間。我總是會選擇適合自己行程的航班，這樣到達後就不會很匆忙了。

N 有道理。但是價格呢？你不覺得這也很重要嗎？

P 當然！價格永遠是一個重要的因素，但我也會考慮到航空公司的聲譽。選擇一家有可靠起降記錄的航空公司非常重要，特別是如果妳有轉機航班的時候。

N 說得好。我也會考慮行李限額以及航空公司是否提供像機上娛樂或餐飲等額外的服務。

P 也別忘了檢查座位安排。有些航空公司提供更大的空間或舒適的座位選擇，這會產生很大的影響，尤其是在長途航班上。

N 這是個好主意。謝謝你的指點，彼特。我想我現在可以做出決定了。

P 很高興能幫上忙。希望妳和妳姊度過愉快的時光。

旅遊字詞補給包

❶ **have a hard time V-ing**　　做……有困難
Anna has a hard time working with her partner.
安娜很難與她的搭檔一起工作。

❷ **upcoming** [ˌʌpˈkʌmɪŋ] *a.* 即將來臨的
I'm really excited about our upcoming trip to Japan.
我對即將到來的日本之旅感到非常興奮。

❸ **be in a rush**　　趕時間
Bill is in a rush, so he can't help you right now.
比爾在趕時間，所以他現在無法幫你。

❹ **reputation** [ˌrɛpjəˈteʃən] *n.* 名聲，名譽
The school has a reputation for being very strict.
那所學校以嚴格聞名。

❺ **reliable** [rɪˈlaɪəbl̩] *a.* 可信賴的，可靠的
Everyone I've spoken to describes Billy as a very reliable young man.
跟我談過的每個人都說比利是個很可靠的年輕人。

❻ **departure** [dɪˈpartʃɚ] *n.* (班機) 起飛
Please arrive at the airport two hours before your departure.
請在班機起飛前兩小時抵達機場。

❼ **baggage allowance** [ˈbæɡɪdʒ əˌlaʊəns]　　行李限額
Make sure to check the baggage allowance before going to the airport.
去機場前記得先確認行李限重。

❽ **make a difference**　　有所不同，有影響
You can make a difference in a child's life by giving him or her encouragement.
多給予孩子鼓勵能使他們的人生有所不同。

經典旅遊實用句

It's important to choose an airline with...
選擇一家有……的航空公司很重要。

用途：說明選擇可靠航空的必要性。

選一家有良好紀錄的航空公司很重要。
It's important to choose an airline with a good track record.
　　　　　　　　　　　　　　　　　an on-time record　準時起降紀錄
　　　　　　　　　　　　　　　　　a safety record　安全紀錄

旅人私房筆記

不同種航班的英文說法

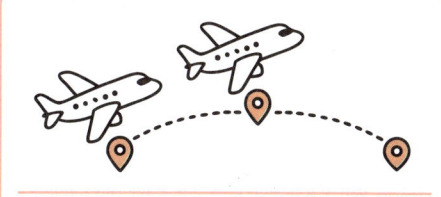

a connecting flight
轉機航班
- 途中需要換班機的航班。

a direct flight
直飛航班
- 途中可能會停其他機場，但乘客不用換班機。

a non-stop flight
直達航班
- 途中完全不落地，直接從出發地飛往目的地。

a red-eye flight
紅眼航班
- 深夜出發，清晨抵達的航班。

Try It Out! 試試看
請依句意在空格內填入適當的字詞

對別人稍微展現一點善意，真的能讓他們的一天有所不同。
Bringing a little kindness to others can really _____ _____ _____ in their day.

Ans make a difference

旅遊英語輕鬆聊
參考答案請掃描使用說明頁上 QR code 下載

Have you ever taken a red-eye flight? How was it?

CH 01 Plan & Preparation · 行前規劃準備

Unit 07 Hotel Reservations
預訂飯店

實境對話 GO！

J Jim（吉姆） **S** Sabina（莎賓娜）

Sabina is calling the Ocean Point Bed & Breakfast to see if she can change her reservation.

J Hello, Ocean Point Bed & Breakfast. This is Jim speaking. How may I help you?

S Hello. I'm calling about **a booking** I **made for** this weekend from Saturday to Sunday. We actually want to stay for two nights. Is there any way I can add a booking for Friday?

J OK, let me check if the room is **available** that day. May I know the last four numbers of the credit card you used to book your room?

S Sure, hold on a second… Yes, it's 3759.

J Great. OK, I see your reservation, and that room is indeed **vacant** the **previous** day.

S Ah, OK.

J I can book it for you right now so that your booking is Friday through Sunday.

S That would be great! Can I ask what the added cost is?

J The **rate** is the same as what you paid for Saturday, so it will be **an extra** NT$2,650 for Friday.

26

- S: OK, and can that be **charged** directly to my card?
- J: Yes, you will receive a second charge from us in that **amount**. Is that OK for you?
- S: Yes, that's fine.
- J: All right. I have just booked the room for another night for you.
- S: Great! Thank you.
- J: Is there anything else I can help you with today?
- S: Ah, yes. I do have one more question about the breakfast. How much does it cost and what time is breakfast served?
- J: The breakfast buffet is included in your room charge. It starts at 6:00 a.m. and ends at 10:00 a.m.
- S: OK, sounds good. Thank you for your help!

> 莎賓娜正打給海角民宿看看是否能更改訂房時間。

- J: 哈囉,這裡是海角民宿。我是吉姆。您需要什麼幫助?
- S: 哈囉。我打來確認我這週末六日訂的房間。我們其實想住兩晚。我能再加訂週五嗎?
- J: 好的,讓我確認那天有無空房。能告訴我您訂房的信用卡號末四碼嗎?
- S: 好的,稍等一下……好了,是 3759。
- J: 很好。好了,我看到您的預約了,那房間前一天剛好也是空的。
- S: 啊,好的。
- J: 我馬上能幫您訂,所以現在您的訂房是從週五到週日。
- S: 那太好了!我能問一下要增加多少費用嗎?
- J: 收費跟您週六住宿的費用一樣,所以會再多收週五的住宿費用新臺幣兩千六百五十元。
- S: 好的,那可以直接用我的卡刷嗎?
- J: 可以,您會收到我們那筆金額的第二次收費通知。這樣您可以嗎?
- S: 可以,很好。
- J: 好。我剛幫您訂好那房間再多住一晚了。
- S: 太好了!謝謝。

> J 今天還有任何需要我幫忙的地方嗎？
> S 啊，對了。我的確還有一個關於早餐的問題。早餐要多少錢及幾點供應呢？
> J 早餐吧包含在您的房間費用內。早上六點開始供應，上午十點結束。
> S 好，聽起來很不錯。感謝你的幫忙！

旅遊字詞補給包

❶ make a booking for... 預訂……
If you want to eat at that restaurant, you'll have to make a booking for a table.
如果你想在那家餐廳用餐，就得先訂桌位。

❷ available [əˋveləb!] *a.* 空的；現有的
Do you have any double rooms available next weekend?
你們下週末還有空的雙人房嗎？

❸ vacant [ˋvekənt] *a.* 空缺的，未被占用的
The hotel still has some vacant rooms for tonight.
這家飯店今晚還有一些空房。

❹ previous [ˋpriviəs] *a.* 以前的，先前的
Tim has forgotten what he learned in the previous chapter.
提姆已經忘了前一章節所學的。

❺ rate [ret] *n.* 費用；價格
The hotel offers a special rate for weekend stays.
這間飯店在週末有特別優惠價格。

❻ an extra + 價錢 / 數字 額外的……（數目）
We need an extra NT$200,000 for the project.
我們這個企劃案另需新臺幣二十萬元才可執行。

❼ charge [tʃɑrdʒ] *vt. & n.* 收費
How much do you charge for delivering the food?
你們食物外送的收費是多少？

❽ amount [əˋmaʊnt] *n.* 數額；數量
The total amount for this purple shirt and that yellow skirt is NT$1,000. 這件紫色的襯衫和那條黃色的裙子總共是新臺幣一千元整。

經典旅遊實用句

Is there any way I can [更動要求]? 　我能再……嗎？

用途：客氣提出修改預約請求。

我能再加訂週五嗎？
Is there any way I can add a booking for Friday?
　　　　　　　　　　　extend my stay　　延長住宿
　　　　　　　　　　　change the date　　更改日期

旅人私房筆記

早餐 buffet 常看到哪些？

- **orange juice** 柳橙汁
- **toast** 烤吐司
- **fried egg** 煎蛋
- **bacon** 培根
- **croissant** 可頌
- **pancakes** 鬆餅
- **milk** 牛奶
- **coffee** 咖啡
- **sausage** 香腸

Try It Out! 試試看

請依句意在空格內填入適當的字詞

首次體驗我們的健身房不收費。
There is no _____ for the first visit to our gym.

Ans　charge

旅遊英語輕鬆聊

參考答案請掃描使用說明頁上 QR code 下載

Imagine you arrived at a hotel one day earlier than your booking. How would you deal with that situation? What would you say to the hotel staff?

Unit 08 Packing for a Trip
打包行李

實境對話 GO！

A Amy（艾咪）　　**J** Jason（傑森）

Jason is making a phone call to his cousin, Amy.

A Hello?

J Hey, Amy. This is Jason.

A Hi, Jason. What's up?

J Well, you've been to Thailand, right?

A Yeah. I went there last summer. Why?

J I'll be visiting there for the first time next month, but I have no idea what I should pack.

A You'll definitely need sunscreen. April is the hottest month in Thailand.

J Oh, wow. I didn't know that. Do you think Bangkok will be too hot?

A It might be. I suggest you plan your day according to the weather. Try to stay indoors during the hottest time of day and go sightseeing outdoors when the sun isn't as intense.

J That's a good idea. What kind of clothing did you wear when you went?

A I mostly wore items made of cotton. As long as what you wear is breathable, you should be comfortable.

J I'm sure I can find some clothes in **a shopping mall** there if I need any more, right?

A Yeah, of course. You'll be able to find everything you need there. Also, all the malls have **air conditioning**, so you can shop around to avoid the heat.

J That's true. Is there anything else you **recommend** bringing?

A Be sure to bring some medicine for motion sickness if you plan on taking any boats. I made the mistake of forgetting last time.

J OK. Let me write everything down.

傑森正在打電話給他的表妹艾咪。

A 喂？

J 嘿，艾咪。我是傑森。

A 嗨，傑森。有什麼事嗎？

J 喔，妳去過泰國對吧？

A 對呀。我去年夏天去的。怎麼了？

J 下個月會是我第一次去那裡玩，但我不知道該帶哪些東西。

A 你一定會需要防晒乳。在泰國，四月是最熱的月分。

J 哦，哇。這我不知道耶。妳覺得曼谷會太熱嗎？

A 應該會。我建議你根據天氣去計劃一天的行程。盡量在一天最熱的時候待在室內，然後在太陽沒那麼大時再到戶外觀光。

J 這是個好主意。妳當時去的時候穿什麼樣的衣服呢？

A 我大多穿棉質的衣服。只要你穿的是透氣的，應該就很舒服。

J 我相信假如我需要更多衣服的話，可以在那裡的購物中心找到的，對吧？

A 當然。你可以在那找到任何你需要的東西。此外，所有的商場都有冷氣，所以你可以逛逛避暑一下。

J 的確是。還有其他建議要帶的東西嗎？

A 如果有打算要坐船的話，務必記得帶些暈船藥。我上次就忘了帶。

J 好的。我來把所有東西都記下來。

旅遊字詞補給包

❶ definitely [ˈdɛfənɪtlɪ] *adv.* 一定，絕對
If we don't hurry up, we'll definitely be late for the concert.
我們若不快一點，演唱會一定會遲到。

❷ sunscreen [ˈsʌnˌskrin] *n.* 防晒乳，防晒油
Don't forget to put on sunscreen before you go to the beach.
去海邊前別忘了擦防晒乳。

❸ sightseeing [ˈsaɪtˌsiɪŋ] *n.* 觀光，遊覽
go sightseeing　　去觀光／遊覽風景
Mia likes to go sightseeing by herself.
米亞喜歡獨自去觀光。

❹ intense [ɪnˈtɛns] *a.* 強烈的，劇烈的
Al suddenly felt an intense pain in his back.
艾爾突然感到背部一陣劇痛。

❺ cotton [ˈkɑtn̩] *n.* 棉；棉花
These clothes are made of pure cotton.
這些衣服是純棉製的。

❻ breathable [ˈbriðəbl̩] *a.* 透氣的
The new shoes are not only comfortable but also breathable.
這雙新鞋不僅舒適，還很透氣。

❼ a shopping mall　　購物中心
The new shopping mall was crowded with shoppers on opening day.
那家新開的購物中心在開幕當天擠滿了購物人潮。

❽ air conditioning [ˈɛr kənˌdɪʃənɪŋ] *n.* 冷氣，空調系統
We can't imagine working without air conditioning during the summer.
我們無法想像夏天工作沒有冷氣。

❾ recommend [ˌrɛkəˈmɛnd] *vt.* 建議；推薦
I recommend this restaurant if you like spicy food.
如果你喜歡吃辣，我推薦這家餐廳。

經典旅遊實用句

Plan your day according to... 你的行程最好根據……來安排。

用途：建議行程要配合天氣。

根據天氣去計劃一天的行程。
Plan your day according to the weather.
　　　　　　　　　　　　　 the temperature　氣溫
　　　　　　　　　　　　　 the forecast　天氣預報

旅人私房筆記

旅遊生病好難受，暈車、暈船、暈機的英文怎麼說？

因快速移動或顛簸引起的頭暈或身體不適就叫 motion sickness（動暈症）。搭乘交通工具而引發的動暈症有：

carsickness
暈車

seasickness
暈船

airsickness
暈機

Try It Out! 試試看　請依句意圈出適當的字詞

❶ We dropped our luggage off at the hotel and went (warning / sightseeing / conditioning).

❷ No one could get near the burning house because of the (intense / initial / inferior) heat.

Ans　❶ sightseeing　❷ intense

旅遊英語輕鬆聊　參考答案請掃描使用說明頁上 QR code 下載

If you were Jason, what three things would you put in your suitcase for Thailand? Why?

33

Notes

Chapter 02 Taking a Flight
搭機

Unit 09
Boarding and Check-in
報到登機

Unit 10
Where's My Passport?!
我的護照呢？！

Unit 11
Going through Security at the Airport　通過機場安檢

Unit 12
Getting Bumped to the Next Flight　候補機位

Unit 13
Flight Delay
班機延誤怎麼辦

Unit 14
Ugh! Flight Canceled!
航班取消真不便！

Unit 15
Missing One's Flight
「機」不可失

Unit 16
Finding Your Seat on the Plane
找座位

Unit 17
A Poorly-Behaved Passenger
飛機上的奧客

Unit 18
In-Flight Service
先生，您還需要什麼呢？

Unit 19
Asking a Flight Attendant for Help　向空姐求助

Unit 20
Last-Minute Gifts　最後才想到買禮物

Unit 21
Getting Directions through the Airport　機場裡找路

Unit 22
Customs Declaration　海關申報

Unit 23
Missing Luggage　消失的行李

Unit 09 Boarding and Check-in
報到登機

S Staff（職員）　　**T** Travis（崔佛斯）

實境對話 GO！

Travis is at the airport **checking in** at an airline counter.

S Hello, can I have your passport and reservation number?

T Of course. Is the flight still **on schedule**?

S Yes, since the weather is **clearing up**, it should **depart** on time.

T Great news. I can't **afford to** be late to my meeting in Shanghai. Is it possible to get a window seat for the flight?

S Let me check. We have a few window seats left at the back of the aircraft. Is that OK?

T Actually, I'll be **in a rush** to get off the plane. Something closer to the front is probably better. An **aisle** seat is fine.

S All right, I will put you in an aisle seat in row seven. Do you have any **luggage** to check?

T Yes, just this one bag.

S I'm sorry. This bag is slightly overweight. We'll have to charge you an extra fee. Would you like to take some heavier items out to avoid paying extra?

T Yes, I have some books in there that I can put in my **carry-on**.

S OK, everything looks fine. Here's your boarding pass. The flight will begin boarding in one hour at Gate 12. It will take about 30 minutes to get through security and immigration.

T Thanks for the heads-up!

S Have a nice flight.

崔佛斯正在機場航空公司的櫃檯辦理登機手續。

S 您好，請問我能看一下您的護照及訂位代碼嗎？

T 那當然。這班班機仍會按預定時間起飛吧？

S 會，由於天氣正逐漸轉晴，班機應該會準時起飛。

T 那真是好消息。我可承擔不起上海的會議遲到。這班次有可能劃到靠窗的位子嗎？

S 讓我幫您看一下。飛機後段還有幾個靠窗的座位。這裡的位子可以嗎？

T 其實我要趕著下飛機。靠越前面的位子可能比較好些。走道的座位也可以。

S 好，那我就把您劃在第七排靠走道的位子。您有行李要託運嗎？

T 有，就這一個袋子。

S 很抱歉。這個袋子有點超重。我們必須向您收取額外的費用。您想拿出一些比較重的物品來避免多付費嗎？

T 好，裡頭有一些書我可以放到隨身包包內。

S 好的，那一切都沒問題了。這是您的登機證。此航班將於一小時後在十二號登機門開始登機。通過安檢及出境檢查會花上約半小時喔。

T 謝謝妳的提醒！

S 祝您旅途愉快。

旅遊字詞補給包

❶ **check-in** [ˈtʃɛk͵ɪn] n. (在機場、旅館、醫院等) 辦理登記手續
Please arrive at the airport two hours before check-in.
請在報到時間前兩小時抵達機場。

❷ **on schedule** 　　　按照預定；按時
ahead of schedule 　　比預定提前 / 超前
behind schedule 　　比預定落後；誤時

The train arrived ahead of schedule. 這班火車比預定時間提早抵達。
Because of the terrorist attacks in the US, all flights were behind schedule.

terrorist [ˈtɛrərɪst] n. 恐怖分子

由於美國發生恐怖攻擊事件，因此所有的航班都延誤了。

❸ **clear up**　（天氣）轉晴，放晴
The rain stopped, and the sky was clearing up.
雨停了，而天空也放晴了。

❹ **depart** [dɪˈpɑrt] vi. 出發，啟程，離開
The travelers gathered on the dock before departing.
遊客們出發前在碼頭集合。

❺ **afford to V**　承擔／負擔得起（做）……
I can't afford to take a month-long trip to Italy.
我負擔不起到義大利度一個月的假。

❻ **in a rush**　趕時間，匆忙地
In modern society, we all seem to be in a rush.
在現今的社會中，大家似乎都匆匆忙忙的。

❼ **aisle** [aɪl] n. (飛機、教堂、戲院的) 走道，過道
I prefer an aisle seat when I fly, so I can stretch my legs.
我搭飛機時喜歡坐靠走道的座位，這樣可以伸展腿部。

❽ **luggage** [ˈlʌgɪdʒ] n. 行李（集合名詞，不可數）
Security checks all the luggage to see if there is anything dangerous inside.　安檢會檢查所有行李，看看裡頭有無任何危險的物品。

❾ **carry-on** [ˈkærɪˌɑn] n. (登機時的) 隨身行李 & a. (指行李) 隨身的，不託運的
My carry-on was too big, so I had to check it in.
我的隨身行李太大了，所以我必須託運。

❿ **boarding pass** [ˈbɔrdɪŋ ˌpæs] n. 登機證
Don't forget to show your boarding pass at the gate.
別忘了在登機門出示你的登機證。

⓫ **security** [sɪˈkjʊrətɪ] n. 保安措施；保全人員
You have to go through security before entering the boarding area.
你必須通過安檢才能進入登機區。

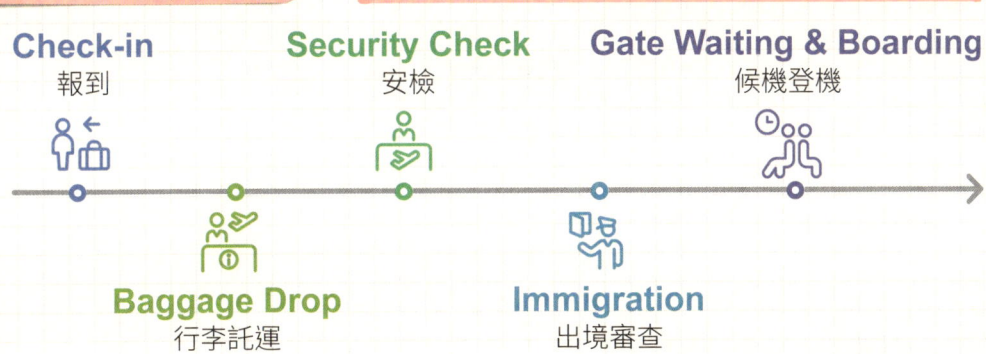

⑫ **immigration** [ˌɪməˈgreʃən] *n.* 移民局出入境檢查；移民（入境）
We had to wait in line at immigration for almost an hour.
我們得在出境檢查處排隊等將近一個小時。

經典旅遊實用句

Can I have your passport and...?
請問我能看一下您的護照及……嗎？

用途：櫃檯確認旅客身分與預約。

請問我能看一下您的護照及訂位代碼嗎？
Can I have your passport and reservation number?
　　　　　　　　　　　　　　　boarding pass　　登機證
　　　　　　　　　　　　　　　booking reference　訂位代碼

旅人私房筆記　　　　　機場報到流程

Check-in 報到
Baggage Drop 行李託運
Security Check 安檢
Immigration 出境審查
Gate Waiting & Boarding 候機登機

Try It Out! 試試看　請依句意在空格內填入適當的字詞

我這次不用託運行李了，因為我只帶了隨身行李。
I don't have to check my luggage this time because I only have a _____.

Ans carry-on

旅遊英語輕鬆聊　參考答案請掃描使用說明頁上 QR code 下載

When checking your luggage at the airport counter, would you try to avoid paying the extra fee, or just pay it?

Unit 10 Where's My Passport?!
我的護照呢？！

實境對話 GO！

C Check-in attendant（櫃檯人員）　　**K** Kyle（凱爾）

Kyle is checking in at the airport when he realizes he has lost his ID.

C Where are you flying today, sir?

K To Chicago, then continuing to New York.

C OK. May I see your passport?

K One second. Oh…It's gotta be here somewhere. Hang on, maybe it's in my bag.

C Take your time.

K Shoot… I think I lost it!

C Stay calm, sir. This happens more often than you would think.

K I've got to make this flight!

C OK, do you have any other form of identification? You can fly within the US without a passport as long as you have some form of ID.

K I have a photocopy of my passport. Oh, and here's my international driver's license.

C Great. Was this a round-trip booking?

K Yeah, I'm just heading home.

C Let me look this up. It says you paid with a credit card. Do you have that card with you?

K That I didn't lose! It's right here.

C Wonderful. We should be able to get you on board. However, you'll **require** an extra **screening** at security.

K Thank you so much!

凱爾在機場辦理登機手續時發現他的證件不見了。

C 先生，您今天要飛往哪裡呢？

K 到芝加哥，然後繼續飛往紐約。

C 好的。我能看看您的護照嗎？

K 我找一下。噢……它一定就在這。等等，說不定是在我包包裡。

C 您慢慢來沒關係。

K 糟了……我想我弄丟護照了！

C 您先別急，先生。這種事情比您想像的還常發生。

K 我一定得搭上這班飛機呀！

C 好，您還有其他的證件嗎？只要您有任何一種身分證明文件，就算沒有護照也能搭乘美國國內航班。

K 我有一張護照的影本。哦，還有這是我的國際駕照。

C 好極了。您訂的是來回機票嗎？

K 是的，我現在是要回家。

C 我來看一下。它上面顯示您是用信用卡付款的。您身上有帶著那張卡嗎？

K 那張我可沒搞丟！就在這呢。

C 太棒了。這樣我們應該就可以讓您登機了。不過，您將需要做一次額外的安全檢查。

K 太感謝妳了！

旅遊字詞補給包

❶ **identification** [aɪˌdɛntəfəˈkeʃən] *n.* 身分證明（文件）（簡稱 ID）
The policeman asked us to show him some identification.
警員要求我們出示證件。

❷ **photocopy** [ˈfotoˌkɑpɪ] *n.* 影印本，複印件
Please hand in a photocopy of your ID card along with the application form.
請在申請表附上一份身分證影本。

❸ **round-trip** [ˌraʊndˈtrɪp] *n.* 往返旅程
I'd like a round-trip ticket from Taipei to Tokyo.
我要一張臺北到東京的來回機票。

❹ **require** [rɪˈkwaɪr] *vt.* 需要；（根據法規）要求，規定
The job doesn't require many people; just a couple of men can do it.
這工作不需很多人做，只要幾個人就可搞定。

❺ **screening** [ˈskrinɪŋ] *n.* 檢查，審查（是否符合要求）
There will be an initial screening of all applicants before the interviews.
　　　　　　　　　→ initial [ɪˈnɪʃəl] *a.* 最初的
在面試前會先對所有應徵者進行初步篩選。

經典旅遊實用句

You'll require [檢查] at security. 您過安檢時會需要……。

用途：告知額外安檢流程。

您過安檢時會需要做一次額外檢查。
You'll require an extra screening at security.
　　　　　　　　an additional check　　額外檢查
　　　　　　　　a manual inspection　　人工檢查

旅人私房筆記

到登機櫃檯該看懂的字！

flight information display
航班資訊顯示螢幕

X-ray machine
X 光機

check-in counter
辦理櫃檯

luggage / suitcase
行李

customer service agent (CSA)
地勤人員

CH 02
Taking a Flight・搭機

Try It Out! 試試看　請依句意在空格內填入適當的字詞

你應該把你的身分證明和旅行文件放在一個方便又安全的地方。
You should keep your ＿＿＿＿＿＿＿＿ and travel documents in a convenient and safe place.

Ans identification / ID

旅遊英語輕鬆聊　參考答案請掃描使用說明頁上 QR code 下載

If you realized you lost your ID while checking in at the airport, what would you do?

43

Unit 11 Going through Security at the Airport
通過機場安檢

C Customs Officer（海關官員）　　**R** Robbie（羅比）

Robbie is waiting in line to get through airport security.

C Hello, sir. Please step forward, and place your carry-on in the basket.

R What should I do with my jacket and the things in my pockets?

C Please use a second basket for all your other **belongings**. But, **hold on to** your passport.

R All right. No problem. Should I take off my shoes?

C No, that won't be necessary. However, liquids and **flammable substances** cannot be taken aboard the plane.

R Oh, I see. Then I guess I should finish drinking this bottle of water.

C Don't take too long; there are lots of people waiting in line, sir. You can throw the bottle into that **bin** over there.

R All right, sorry to **slow** things **down**.

C When you are ready, please walk slowly through the body scanner. After your bags make it through the X-ray screening machine, you can pick them up on the other side.

R I should **mention** that I have a metal **implant** in my shoulder. It always **sets off** metal detectors.

C Don't worry, sir. An officer will give it a check.

R Thank you, officer.

C Thank you, and have a safe flight.

(Robbie walks through the body scanner and goes to get his things.)

羅比正排隊等候通過機場安檢。

C 先生，您好。請往前站，然後將您的隨身行李放置於籃子裡。

R 請問我的夾克跟口袋內的物品要怎麼辦？

C 請用另一個籃子放您所有其他的隨身物品。但護照請您自己留著。

R 好的。沒問題。我需要脫鞋嗎？

C 不用，沒那個必要。但是液體及易燃物質不能帶上飛機。

R 喔，了解。那我想我應該要喝完這瓶水。

C 那請您盡快，先生；後面還有很多人在排隊。您可以把瓶子丟到那邊的垃圾桶裡。

R 好的，不好意思拖延到進度。

C 您準備好時就請慢慢走過人體掃描機。您的袋子通過 X 光掃描機後，就可以到另一邊去領取。

R 我得提一下我的肩膀有金屬移植物。它每次都會讓金屬探測器作響。

C 先生，不用擔心。會有一位人員為您檢查。

R 謝謝您。

C 感謝，祝您旅途平安。

(羅比走過人體掃描機，然後前去領取他的物品。)

旅遊字詞補給包

❶ **belongings** [bəˈlɔŋɪŋz] *n.* 攜帶物品；所有物 (恆為複數)
Martha led a simple life and had few belongings.
瑪莎過著擁有很少物品的簡樸生活。

❷ **hold on to...**　　保留 / 留著……；抓住 / 緊握……
The mother held on to her little girl while they walked through the crowd of people.
那位母親緊抓著她的小女兒，在那一群人中間穿梭而過。

❸ **flammable** [ˈflæməbl̩] *a.* 易燃的
Keep flammable materials away from the stove.
易燃物品要遠離火爐。

❹ **substance** [ˈsʌbstəns] *n.* 物質
Many natural substances are used in Chinese medicine.
許多天然物質都會用在中藥裡。

❺ **bin** [bɪn] *n.* 垃圾桶；箱子
Please throw the paper into the bin.
請把這張紙丟進垃圾桶。

❻ **slow... down / slow down...**　　使……慢下來
slow down　　慢下來
Please slow down; you've been driving too fast.
請慢下來；你車開得太快了。

❼ **mention** [ˈmɛnʃən] *vt.* 提及
Lucy mentioned people who most supported her when she accepted the award.
露西領取獎項時，提到了那些最大力支持她的人。

❽ **implant** [ˈɪmplænt] *n.* 植入物
Jack needs a dental implant to replace his missing tooth.
傑克需要植牙來取代掉了的那顆牙。

❾ **set off... / set... off**　　觸發 (警報)；燃放 (煙火、鞭炮)
Kevin bought a lot of fireworks and set them off during the Fourth of July celebrations.
凱文買了很多煙火，在美國國慶日的慶祝活動中燃放。

fireworks [ˈfaɪrˌwɝks] *n.* 煙火 (恆用複數)

經典旅遊實用句

Please use a second basket for all your [物品].
請用另一個籃子放您所有的……。

用途：要求分開放物品。

請用另一個籃子放您所有其他的隨身物品。
Please use a second basket for all your <u>other belongings</u>.
　　　　　　　　　　　　　　　　　　<u>electronics</u>　　電子產品
　　　　　　　　　　　　　　　　　　<u>personal items</u>　個人物品
　　　　　　　　　　　　　　　　　　<u>loose change</u>　　零錢

旅人私房筆記　　　　　　　　安檢小字典

security [sɪˋkjʊrətɪ]
n. 安檢；安全 (措施，不可數)

body scanner
[ˋbɑdɪ ˏskænɚ]
n. 人體掃描機

metal detector
[dɪˋtɛktɚ]
n. 金屬探測器

carry-on [ˋkærɪˏɑn]
a. 可隨身攜帶的 &
n. 隨身行李

X-ray screening
[ˋskrinɪŋ] **machine**
X 光掃描機

Try It Out! 試試看　　請依句意在空格內填入適當的字詞

下火車時務必要帶好個人物品。
Be sure to take your personal _____ when you get off the train.

Ans belongings

旅遊英語輕鬆聊　　參考答案請掃描使用說明頁上 QR code 下載

Do you usually prepare well before going through airport security? What do you do to get ready?

CH 02　Taking a Flight・搭機

47

Unit 12 Getting Bumped to the Next Flight
候補機位

實境對話 GO！

A Attendant（櫃檯人員）　　**K** Kevin（凱文）

Kevin is waiting to board his flight at the gate.

A Would passengers, Kevin Liu, Hiro Tanaka, and Jane Smith, please come to the counter?

K Hello, I'm Kevin Liu. What is it?

A Sir, this flight is currently overbooked. Since you're flying standby, we're going to have to put you on the next flight which is in six hours.

K But I have a connection that I have to make in Tokyo.

A Are you also booked with us for that flight?

K No, I think I'm with Air Japan.

A That's fine. They're one of our partner airlines. Let me look up the flight details, and we'll see what we can do.

K OK, thank you very much.

A Well, I'll change you to another one of our partners, so you should be able to make your transfer with time to spare.

K That would be great! When is that flight?

A It's **scheduled** for two hours from now over at gate D26. It looks like the flight is fairly empty, so we can bump you up to first class if you'd like.

K **Absolutely**! Thanks a million.

凱文正在登機門等待登機。

A 旅客劉凱文、田中廣和珍・史密斯，請至櫃檯。

K 哈囉，我是劉凱文。有什麼事嗎？

A 先生，目前該航班座位超訂。由於您是候補的乘客，我們將把您安排到六小時後的下一班次。

K 可是我必須在東京轉機。

A 那班航班也是訂我們航空公司的嗎？

K 不是，好像是日本航空。

A 沒問題。他們是我們的聯盟航空之一。讓我查一下航班細節，我們再看看我們能怎麼做。

K 好的，非常感謝妳。

A 嗯，我將您換到另一個聯盟航空的班機，所以您應該會有足夠的時間可以轉機。

K 那真是太好了！班機是什麼時候呢？

A 預計時間在兩小時後的 D26 登機門。看來那班航班還有相當多的空位，所以如果您願意的話，我們可以讓您升級到頭等艙。

K 那當然！太感謝了！

旅遊字詞補給包

❶ **attendant** [ə`tɛndənt] *n.* 服務人員
The flight attendant asked us to fasten our seat belts.
空服員請我們繫好安全帶。

❷ **currently** [`kɝəntlɪ] *adv.* 現今，目前
There are many foreign students currently studying in Taiwan.
現今有不少外國學生在臺灣求學。

❸ **connection** [kəˋnɛkʃən] *n.* 銜接／接駁交通工具(如飛機、火車和公車等)
Jewel's train got delayed, so she missed her connection to Taitung.
茱兒的火車誤點，所以她錯過了去臺東的銜接列車。

❹ **partner** [ˋpɑrtnɚ] *n.* 合夥人
Germany is one of our major trading partners in Europe.
德國是我們在歐洲主要的貿易夥伴國之一。

❺ **time to spare** 多出的時間
spare [spɛr] *vt.* 抽出，騰出 (時間、金錢或人手)
Roger isn't in a hurry to get to the interview, because he still has time to spare.
羅傑不用趕著去面試，因為他還有多餘的時間。

單數：passerby [ˋpæsɚˏbaɪ] *n.* (過) 路人

The beggar asked passersby to spare him some change.
那個乞丐請過路人給他一些零錢。

❻ **schedule** [ˋskɛdʒʊl] *vt.* 排訂，將……安排在
The opening of the department store is scheduled to take place this Sunday.
百貨公司預定在這個星期天開始營業。

❼ **absolutely** [ˋæbsəˏlutlɪ] *adv.* 當然 (口語用法)
A: Would you like to go to the prom with me?
B: Absolutely!
甲：妳願意跟我一起參加舞會嗎？
乙：當然！

經典旅遊實用句

We can... if you'd like. 如果您願意的話，我們可以……。

用途：提供補償。

如果您願意的話，我們可以讓您升級到頭等艙。
We can bump you up to first class if you'd like.
　　　　upgrade your seat　　升等您的座位
　　　　move you to business class　　將您換到商務艙

旅人私房筆記 — 認識飛機艙等

First Class
頭等艙（F）

Premium Economy
豪華經濟艙（W）

Business Class
商務艙（C/J）

Economy Class
經濟艙（Y）

Try It Out! 試試看 — 請依句意選出適當的字詞

The movie was _____ for release in January next year.
- Ⓐ scheduled
- Ⓑ overbooked
- Ⓒ transferred
- Ⓓ bumped

Ans Ⓐ

旅遊英語輕鬆聊 — 參考答案請掃描使用說明頁上 QR code 下載

If you got a free upgrade to first class, what would you do first when you got on the plane? Why?

CH 02 Taking a Flight・搭機

Unit 13 Flight Delay
班機延誤怎麼辦

實境對話 GO！

L Louis（路易斯）　　**C** Charlotte（夏綠蒂）

Louis is waiting for his flight. Charlotte, one of the plane's crew members, *informs* him that the flight got delayed.

L Excuse me, I was wondering why we haven't started **boarding** yet. It's less than an hour until **takeoff**.

C I think you may have missed the **announcement**. This flight has been delayed until 10:30 p.m. tonight.

L What?! But I need to make this connecting flight in order to be in Melbourne by morning. There's no way I'll get there in time with this long delay.

C I'm very sorry, sir. There is a mechanical problem with the plane that will need several hours to get fixed.

L I can't believe this!

C Sir, the most important thing is that we get you to where you're going safely, even if that means being late.

L I get that, but what **am I supposed to** do for the next 12 hours?

C Sadly, it seems the only thing to do is to wait for the plane to be ready.

L For 12 hours?! Don't you give out **vouchers** for food and drinks at the restaurants here? Or put up your passengers at the airport hotel? Twelve hours is a long time to wait at the airport.

C I'm very sorry, sir. As we are a **budget airline**, we don't provide any of those **perks**.

L Well, this is just quite inconvenient. Are there even any other choices?

C You can always stay at the airport lounge. That may be a more **appealing** choice than waiting here at the gate.

L Yeah, but the airport **lounge** is costly. I guess I'm just going to spend the next 12 hours here complaining about this delay on Twitter!

路易斯正在等候他的班機。班機的其中一位機組人員──夏綠蒂，向他通知班機延誤了。

L 不好意思，我想請問一下為什麼還沒開始登機？現在離起飛時間已經不到一小時了。

C 我想您大概是錯過了廣播宣布。這個航班會延誤到今晚十點三十分。

L 什麼？！可是我得搭上這架轉接班機，才能在明天早上抵達墨爾本。延誤這麼久，我絕對沒辦法及時趕到。

C 先生，非常抱歉。飛機有機械方面的問題，要花好幾個小時才能修復。

L 我真不敢相信！

C 先生，最重要的一點是我們會安全地帶您到達目的地，即便這意味著會晚到。

L 這我懂，不過接下來的十二個小時我應該做什麼才好？

C 不幸的是，看來唯一能做的事就只有等待飛機準備就緒。

L 等十二個小時？！你們不是會發放這邊餐廳的食物和飲料抵用券嗎？或是安排乘客到機場旅館住宿？在機場裡等十二小時可是一段很漫長的時間。

C 先生，很抱歉。由於我們是廉航，因此我們不會提供任何這些福利。

L 嗯，這樣真是相當不方便。那有任何其他選擇嗎？

C 您隨時可以去機場休息室休息。相較於在登機門前等候，那或許是一個比較吸引人的選項。

L 是沒錯，不過機場休息室很貴。我猜接下來的十二個小時，我只好待在這裡在推特上發誤點的抱怨文！

旅遊字詞補給包

❶ inform [ɪnˋfɔrm] *vt.* 通知，告知
A notice informing the staff about the policy changes was posted.
公司貼出告示，告知全體員工一些公司政策上的變動。

❷ board [bɔrd] *vi. & vt.* 登上 (船、火車或飛機)
Passengers to Houston, Texas, please board the train at Platform 16.
前往德州休士頓的旅客，請至十六號月臺登車。

❸ takeoff [ˋtek͵ɔf] *n.* 起飛，升空
We waited for the takeoff announcement. 我們等待起飛廣播。

❹ announcement [əˋnaʊnsmənt] *n.* 宣告，宣布
announce [əˋnaʊns] *vt.* 宣布，公告
Our boss made an unexpected announcement this morning.
今天早上我們老闆發表了一項出人意料的聲明。
My sister surprised everyone by announcing she was getting married.
我妹妹宣布要結婚，把所有人都嚇了一跳。

❺ be supposed to V 應該 (做)……
Ricky is late. He was supposed to meet me here at four o'clock.
瑞奇遲到了。他應當要在四點鐘在這裡跟我碰面的。

❻ voucher [ˋvaʊtʃɚ] *n.* 抵用券；代金券；優惠券
You can use this voucher to get a free drink.
你可以用這張抵用券換一杯免費飲料。

❼ budget airline [ˋbʌdʒɪt ˋɛrlaɪn] *n.* 廉價航空 / 廉航
We booked our trip with a budget airline to save money.
我們訂廉航機票以節省旅費。

❽ perk [pɝk] *n.* 額外待遇 / 收入；津貼
One of the perks of working here is a free lunch every day.
在這裡工作的福利之一是每天都可以吃免費午餐。

❾ appealing [əˋpilɪŋ] *a.* 有吸引力的，誘人的
The idea of not having to get up early every day is rather appealing to Annie. 對安妮來說，不需要每天早起是個十分誘人的主意。

❿ lounge [laʊndʒ] *n.* 休息室；會客廳
We waited for our flight in the airport lounge.
我們在機場貴賓室等候航班。

經典旅遊實用句

I was wondering why... 我想請問一下為什麼……？

用途：禮貌詢問原因。

我想請問一下為什麼還沒開始登機？
I was wondering why we haven't started boarding yet.
　　　　　　　　 the gate is still closed　　登機門還沒開
　　　　　　　　 there's no announcement　　沒有廣播通知

旅人私房筆記　　　廉價航空 vs 傳統航空

budget airline 廉價航空		full-service airline 傳統航空
便宜	票　價	較高
擁擠	座位空間	寬敞
需額外付費	行李政策	通常包含行李
需付費	餐飲服務	免費餐飲
無餐食或住宿	航班延誤處理	提供餐食或住宿

Try It Out! 試試看　　請將下列字詞組合成一正確通順的句子

These _____
(last / batteries / supposed to / are) for a year.

　　　　　　　　　　　　Ans　batteries are supposed to last

旅遊英語輕鬆聊　　參考答案請掃描使用說明頁上 QR code 下載

What would you do if you had to wait for twelve hours at the airport without any meals or entertainment?

Unit 14
Ugh! Flight Canceled!
航班取消真不便！

實境對話 GO！

I Isaac（艾薩克）　**G** Gina（吉娜）

Isaac just arrived at the airport and found out his flight was *canceled*. He asks help from Gina, who is one of the airline's ground staff members.

I Excuse me, I just received notice that my flight has been canceled due to the typhoon. Can you tell me what's going on?

G I'm sorry for the inconvenience, but due to the typhoon, several flights have been canceled for safety reasons.

I Oh, no. Do you know how long the typhoon will last?

G It's hard to say for sure, but they're expecting heavy rain and strong winds for the next 24 hours.

I I see. Is there any chance I can get on a flight tomorrow or the day after tomorrow?

G Let me check the available flights for you. Could you please provide me with your booking number and passport?

I Sure, here they are.

G It looks like we have a few flights available for tomorrow. Would you like me to book one for you?

I Yes, please. Thank you for your help.

G It's the least I can do. I've rescheduled you on the 10 p.m. flight to Toronto tomorrow. Is this acceptable for you?

I Yes, that's perfect.
G Please pay attention to any notices regarding your new flight **in case** the weather gets worse. If you have any more questions, I'm happy to help.
I I will. Thank you.

艾薩克剛抵達機場，發現自己的班機被取消了。他向吉娜尋求協助，她是航空公司的一名地勤人員。

I 不好意思，我剛剛收到通知，我的班機因颱風被取消了。妳能告訴我發生了什麼事嗎？
G 造成您的不便，我深感抱歉，由於受颱風的影響，出於安全考量，多架班機被取消了。
I 哦，不。妳知道颱風會持續多久嗎？
G 這很難確定，但預計未來二十四小時會有強風豪雨。
I 我明白了。我有機會搭乘明天或後天的班機嗎？
G 讓我為您查一下可預訂的班機。可以給我您的訂位號碼和護照嗎？
I 當然，在這裡。
G 看來我們明天還有幾架班機可供訂位。您需要我為您預訂一班嗎？
I 好，麻煩妳。謝謝妳的幫助。
G 舉手之勞而已。我已為您重新安排搭乘明天晚上十點飛往多倫多的班機。您可以接受嗎？
I 可以，太完美了。
G 請留意任何有關您新班機的任何通知，以防天氣惡化。如果您還有其他問題，我很樂意協助。
I 我會的。謝謝。

旅遊字詞補給包

1. **cancel** [ˈkænsl̩] *vt.* 取消
   We canceled the baseball game because of the rain.
   因為下雨，我們取消了棒球比賽。

❷ **typhoon** [taɪˈfun] *n.* 颱風
The typhoon damaged the whole village.
颱風摧毀了整個村莊。

❸ **available** [əˈveləbl̩] *a.* 可提供的，可得到的
The store offers a wide range of shirts available in different sizes and colors.
這家商店提供各種不同尺寸和顏色的襯衫。

❹ **booking** [ˈbʊkɪŋ] *n.* 預訂
book [bʊk] *vt.* 預訂 (座位、門票等)
I'd like to make a booking for a table for two this afternoon.
我想預訂今天下午的兩人桌位。
I'd like to book a train ticket to Taipei for next Friday.
我要訂一張下週五去臺北的火車票。

❺ **acceptable** [əkˈsɛptəbl̩] *a.* 可接受的
Wearing jeans to the formal party is not acceptable.
穿牛仔褲參加正式宴會是不被接受的。

❻ **in case...**　　以防 (萬一) ……
You should bring an umbrella in case it rains this afternoon.
你應該帶把雨傘，以防今天下午下雨。

經典旅遊實用句

Please pay attention to any notices regarding your...
請留意有關您……的任何通知。

用途：提醒旅客後續注意通知 (天氣變動或改登機門)。
請留意任何有關您新班機的任何通知。
Please pay attention to any notices regarding your new flight.

gate	登機門
boarding time	登機時間
flight status	航班狀態

旅人私房筆記

人到機場卻看不懂？機場重要資訊的英文單字

每天有數以萬計的旅客出國，如果看不懂機場標示或是資訊，很容易在機場迷路甚至是錯過班機！以下是幾個常在機場看到的重要英文單字，一起看看吧：

canceled	取消	**gate**	登機門
on time	準時	**terminal**	航廈
delayed	延遲	**arrivals**	入境大廳
boarding	登機	**departures**	出境大廳

CH 02　Taking a Flight・搭機

Try It Out! 試試看　　請依句意在空格內填入適當的字詞

① 由於颱風，學校決定取消當天的課程。
The school decided to _____ classes for the day because of the typhoon.

② 在包包裡放一些零食，以防你等一下餓了。
Pack some snacks in your bag _____ _____ you get hungry later.

Ans　① cancel　② in case

旅遊英語輕鬆聊　　參考答案請掃描使用說明頁上 QR code 下載

If you have no choice but to sleep at the airport overnight, where would you sleep?

Unit 15 Missing One's Flight
「機」不可失

實境對話 GO!

A Anita（艾妮塔）　　E Employee（服務人員）

Anita is at the airport talking to an airline employee.

A Excuse me, is the 5:30 flight to Vancouver still boarding? My cab was involved in an accident. I know I'm terribly late.

E It is already 5:20 p.m. They completed boarding about five minutes ago.

A Can you stop the flight for me and let me get on? I really can't miss it.

E I'm sorry, ma'am. I believe the plane is pulling away from the gate and moving towards the runway.

A Oh no! You mean I can't make it? This is horrible.

E I'm sorry, miss. If you had arrived just 20 minutes earlier, we might have been able to rush you to the gate, but you are too late.

A What can I do? Please help me out. Today is not my day.

E The most we can do is switch your reservation to the next flight bound for Vancouver. However, it won't be taking off for another six hours.

A That's fine. Please give me any seat that is available. Is there an extra fare to change flights?

E Yes, there is a small fee. Let me see your passport and booking information. We'll do our best to get you on the 11:45 p.m. flight.

> 艾妮塔在機場與航空公司人員談話。

A 不好意思，五點半飛往溫哥華的班機還在登機中嗎？我搭的計程車碰到了事故。我知道我太晚到了。

E 現在已是下午五點二十分。大概五分鐘前就已經登機完畢了。

A 你可以幫我擋下飛機，然後讓我登機嗎？我真的不能錯過這架航班。

E 這位女士，很抱歉。我想飛機現在正駛離登機口、朝跑道前進了。

A 不會吧！你的意思是我搭不上飛機了嗎？真是糟糕。

E 小姐，很不好意思。如果您早二十分鐘到，我們還可能可以趕緊送您到登機口，但您真的太晚到了。

A 我能怎麼辦？請幫幫我。我今天真倒楣。

E 我們最多能幫您把您的訂位換到下一班飛往溫哥華的班機。不過這架班次六小時後才起飛。

A 沒關係。麻煩有任何空位就給我吧。更改航班有額外的費用嗎？

E 是的，會有一小筆費用。讓我看看您的護照和訂位資料。我們會竭盡所能幫您搭上晚上十一點四十五分的班機。

旅遊字詞補給包

❶ **employee** [ˌɪmplɔɪˈi] *n.* 員工，受僱者
The firm provides its employees with many benefits.
那間公司提供旗下員工很多的福利。

❷ **be involved in...**　　涉入……
involve [ɪnˈvɑlv] *vt.* 涉及，牽涉；包含
The evidence shows that Freddy was involved in that murder.
證據顯示弗萊迪涉入那起命案。

❸ **pull away**　　（車輛、飛機）駛離，開出
The bus pulled away from the main station around noon.
那輛公車於中午左右駛離總站。

❹ **horrible** [ˈhɔrəbḷ] *a.* 糟糕的；可怕的
Traffic there is horrible this time of day. It is better if we go later.
每天此時那裡的交通都很糟糕。我們最好晚點再去。

❺ **rush sb/sth to...**　　將某人／物急速送往……
rush [rʌʃ] *vt.* 快速運輸；速送
The ambulance rushed the injured person to the hospital.
那輛救護車把傷患火速送往醫院。

❻ **switch** [swɪtʃ] *vt. & vi.* 轉換，變更
Mark used to vote for that party, but he is switching his vote to a different one this year.
馬克以前都投票給該黨，但他今年要轉投給不同黨派。

❼ **bound for...**　　前／開往……
bound [baʊnd] *a.* 去……的；準備前往……的
I'm afraid you've taken the wrong train. This train is bound for New York, not Chicago.
你恐怕搭錯了火車。這班車是開往紐約，並非芝加哥。

❽ **fare** [fɛr] *n.* (交通工具的) 費用；車費
How much is the MRT fare from Shilin to Taipei Main Station?
從士林到臺北火車站的捷運車資是多少？

經典旅遊實用句

My cab was involved in [事件]. I know I'm terribly late.
我搭的計程車遇上了……。我知道我太晚到了。

用途：解釋自己遲到的具體原因。

我搭的計程車碰到了<u>事故</u>。我知道我太晚到了。
My cab was involved in <u>an accident</u>. I know I'm terribly late.

　　　　　　　　　　a traffic jam　　塞車
　　　　　　　　　　a flat tire　　　爆胎

62

旅人私房筆記

看懂登機證

BOARDING PASS — BUSINESS CLASS — **BOARDING PASS**

NAME OF PASSENGER 旅客姓名	FLIGHT 航班號碼	NAME OF PASSENGER
JOE BROWN	F3924	JOE BROWN
FROM: 出發地	SEAT 座位	FROM:
DENPASAR	A18	DENPASAR
TO: 目的地	BUSINESS CLASS 商務艙	TO:
LONDON		LONDON
DATE 日期 / TIME 起飛時間 / GATE 登機口	BOARDING TIME 登機時間	FLIGHT F3924 / SEAT A18
12 MAY / 14:30 / 3	13:50	

BUSINESS CLASS — BUSINESS CLASS

CH 02 — Taking a Flight · 搭機

Try It Out! 試試看
請依句意在空格內填入適當的字詞

❶ 自從約翰買了筆記型電腦，他就把玩電動的時間改到晚上。
Ever since John bought a laptop computer, he has _____ to playing video games in the evening.

❷ 我們公司剛成立時只有兩名員工。
When our company first started out, there were only two _____.

Ans ❶ switched ❷ employees

旅遊英語輕鬆聊
參考答案請掃描使用說明頁上 QR code 下載

If you could give Anita one piece of advice, what would you tell her?

63

Unit 16 Finding Your Seat on the Plane

找座位

F Flight attendant（空服員）　　**L** Leonard（雷納德）

Leonard boards the plane and is looking for his seat.

F Welcome aboard our nonstop flight to Johannesburg.

L Hello. Can you **direct** me **to** my seat? I'm not sure which side of the plane it's on.

F Certainly. May I see your boarding pass?

L Here you go. Sorry, I'm not used to flying, and this is a really big plane.

F You're in row 21, seat F. That will be down this aisle on the right side. You can put your carry-on in the bin **overhead**.

L Thank you very much.

(Leonard heads down the aisle, **only to** find someone sitting in seat 21F.)

L Excuse me, but there seems to be someone in my seat. Is my ticket wrong?

F Oh, I see. Give me a second to check their ticket.

(The flight attendant returns after **sorting out** the situation.)

F Sorry for the **misunderstanding**. Your seat is 21F, so this is the right seat. The other passenger **misread** their boarding pass. They will be **seated** in row 27F.

L Thank you so much. Sorry for the trouble.

F It's no problem at all. Please enjoy the flight. Let me know if there is anything else I can help with.

L Actually, could I get an extra blanket in case it gets a bit **chilly**?

雷納德登機後在找他的位子。

F 歡迎搭乘我們飛往約翰尼斯堡的直達班機。

L 哈囉。請問妳能告訴我我的位子在哪嗎？我不太確定是在飛機的哪一邊。

F 當然沒問題。我能看一下您的登機證嗎？

L 給妳。抱歉，我不習慣搭飛機，而且這架飛機真大。

F 您在第二十一排、座位 F。這條走道直走下去，然後在您右手邊。您可以把隨身行李放置在頭上的置物箱。

L 謝謝妳。

（雷納德穿越走道，卻發現有人坐在座位 21F。）

L 不好意思，但似乎有人坐了我的位子。是我的機票錯了嗎？

F 喔，我了解了。給我一點時間，我去看看那個人的機票。

（空服員處理好狀況後回來。）

F 抱歉誤會一場。您的位子是 21F，是正確的位子沒錯。另一名乘客看錯了登機證。他會坐在 27F。

L 謝謝妳。很抱歉麻煩妳了。

F 一點也不麻煩。祝您旅途愉快。如果有什麼需要我幫忙的地方，隨時可以告訴我。

L 事實上，可以再多給我一件毯子以免變得有點冷嗎？

旅遊字詞補給包

① **direct sb to +** 地方　　指引某人到某地
Could you direct me to the campus bookstore?
可不可以請你跟我說學校書店該怎麼去？

② **overhead** [ˋovɚˏhɛd] *adv. & a.* 在頭頂上（的）；在空中（的）
Teddy saw a flock of birds flying overhead on their way south for the winter.　　泰迪看到一群要到南方過冬的鳥兒從上空飛過。

65

❸ **S + V, only to V**　　結果卻 / 竟然……
Robert looked for his keys everywhere, only to remember he had left them in the car.
羅伯特到處找鑰匙，結果才想起自己忘在車裡了。

❹ **sort out... / sort...out**　　處理好 / 解決……；整理 / 分類……
Look at the mess in your room. You need to sort it out before lunch.
看看你房間一團亂。你必須在午餐前把它整理好。

❺ **misunderstanding** [ˌmɪsʌndɚˈstændɪŋ] *n.* 誤會
To avoid any possible misunderstanding, we should have a face-to-face meeting.
為避免任何可能的誤會，我們應該當面開會。

　　　　　　　　　　　　　　　　　　　　　三態同形

❻ **misread** [ˌmɪsˈrid] *vt.* 讀錯；看錯；誤解
Emma misread her friend's email and ended up at the wrong meeting place.
艾瑪把她朋友的電子郵件看錯，結果跑到錯的會面地點。

❼ **seated** [ˈsitɪd] *a.* 入座的，坐下來的
Everyone must be seated before the plane takes off.
飛機起飛前，每個人都要在位子上坐好。

❽ **chilly** [ˈtʃɪlɪ] *a.* 寒冷的
Nothing can compare to enjoying a hot spring on a chilly day.
沒有什麼比在冷天泡溫泉更享受的了。

經典旅遊實用句

Could I get an extra [物品]?　　我可以再要一個……嗎？

用途：索取額外物品。

可以再多給我一件毯子嗎？
Could I get an extra blanket?
　　　　　　　　　　pillow　　　　枕頭
　　　　　　　　　　headset　　　　耳機
　　　　　　　　　　bottle of water　一瓶水

旅人私房筆記

圖解機艙大小事

overhead bin
頭頂置物櫃

life vest
救生衣

seat belt
安全帶

aisle seat
靠走道座位

middle seat
中間座位

window seat
靠窗座位

CH 02 Taking a Flight．搭機

Try It Out! 試試看
請依句意在空格內填入適當的字詞

唐納去找老師以試著解決他的數學問題。
Donald went to his teacher to try to _____ _____ his math problems.

Ans sort out

旅遊英語輕鬆聊
參考答案請掃描使用說明頁上 QR code 下載

> If you could sit anywhere on the plane, which seat would you choose and why?

67

Unit 17 A Poorly-Behaved Passenger
飛機上的奧客

K Kevin（凱文）　　**P** Passenger（旅客）　　**F** Flight attendant（空服員）

Kevin walks to his seat to find someone already there.

K Excuse me. I believe that this is my seat.

P You're wrong. Buzz off and find your own place to sit.

K This is 3A, right?

P Who cares if it is?

K This is the seat on my ticket.

P That must be your mistake. You don't **belong** in first class. How could it be your seat, huh?

(Kevin goes to find a flight attendant.)

K **Pardon me**, but there's a man sitting in my seat, and he won't leave.

F Are you sure it's your seat?

K I'm **positive**. My ticket clearly says seat 3A.

F The larger man over there?

K Yep. That's the guy.

F He looks **pretty drunk**. Wait here.

(The flight attendant goes to talk to the man.)

F Sir, could you please show me your ticket?

P I don't know. Could you show me your ticket?
F If you don't have one, you will be asked to leave the plane.
P I'm not going anywhere!
(Later, airport police take the drunk passenger off of the plane.)
F I'm sorry about that. Let me know if there's anything we can do for you.
K Thanks again!

凱文走到他的座位卻發現位子上已經有人了。

K 不好意思。我想這個座位是我的。
P 你錯了。滾開,去找你自己的座位坐。
K 這裡是 3A,對吧?
P 誰管它是不是啊?
K 這就是我機票上的座位。
P 那一定是你搞錯啦。你不配坐頭等艙。這怎麼會是你的座位呢,蛤?
(凱文跑去找空服員。)
K 不好意思,有個人坐在我的座位上,而且他不離開。
F 您確定那是您的座位嗎?
K 絕對是。我的機票上清清楚楚寫的是座位 3A。
F 是那邊那個壯漢嗎?
K 對。就是他。
F 他看起來相當醉。您在這兒等一下。
(空服員過去和那個人談話。)
F 先生,能請您把您的票給我看一下嗎?
P 不知道啦。那妳能給我看一下妳的票嗎?
F 如果您沒有機票的話,那您會被要求離開這架班機。
P 我哪都不去!
(後來,航警把酒醉的乘客帶下飛機。)
F 對此我很抱歉。如果有什麼我們可以為您效勞的請告訴我。
K 再次感謝!

旅遊字詞補給包

❶ **-behaved** [bɪˈhevd] *suffix* 表現……的，行為……的
My brother used to be a naughty boy, but now he's a well-behaved young man.
我弟弟曾經是個頑皮的孩子，但現在已經是個行為端正的年輕人了。

❷ **belong** [bɪˈlɔŋ] *vi.* 屬於，應在 (某處)
This long table doesn't belong in this room. It goes in the dining room.
這張長桌不應被放在這個房間內。它應該擺在飯廳裡。

❸ **pardon me** 不好意思，對不起；請再說一遍
Pardon me, could you say that again?
不好意思，你可以再說一次嗎？

❹ **positive** [ˈpɑzətɪv] *a.* 肯定的，有把握的；正面的；積極樂觀的
I am pretty positive that Elva will be on my side of this argument.
我相當肯定在這次爭論中，艾娃會站在我這邊。

❺ **pretty** [ˈprɪtɪ] *adv.* 相當地 & *a.* 漂亮的
I'm pretty tired after the long flight.
這趟長途飛行後我相當累。

❻ **drunk** [drʌŋk] *a.* 喝醉的
Steve was so drunk last night that he could barely stand up.
史蒂夫昨晚醉到幾乎站不起來。

經典旅遊實用句

He looks pretty…. 他看起來相當……。

用途：描述對方狀況不佳。

他看起來相當<u>醉</u>。
He looks pretty <u>drunk</u>.
　　　　　　<u>tired</u>　　　　疲累
　　　　　　<u>aggressive</u>　很兇
　　　　　　<u>confused</u>　　困惑

旅人私房筆記　看懂飛機上的標誌

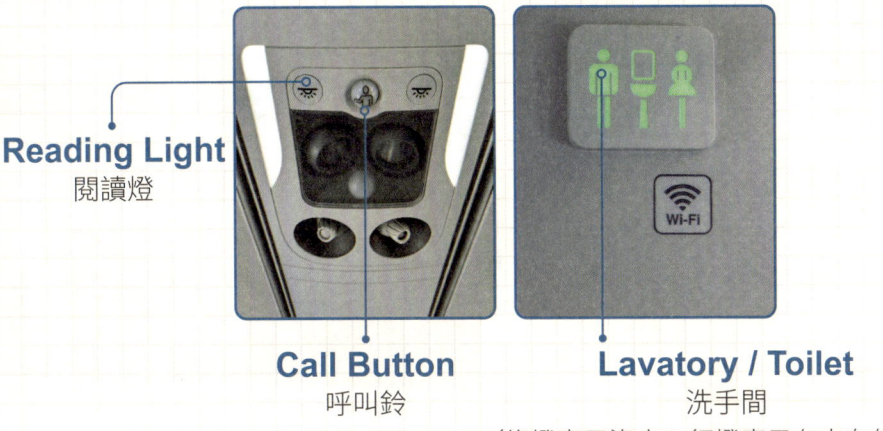

Reading Light
閱讀燈

Call Button
呼叫鈴

Lavatory / Toilet
洗手間
（綠燈表示沒人；紅燈表示有人在使用中）

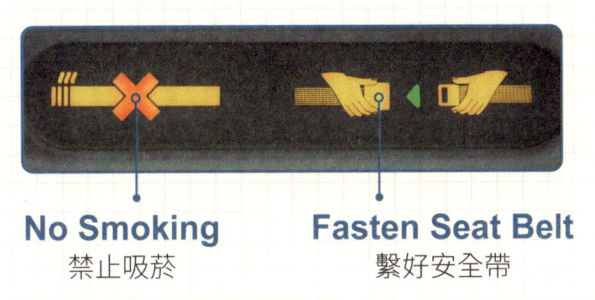

No Smoking
禁止吸菸

Fasten Seat Belt
繫好安全帶

Try It Out! 試試看　請依句意在空格內填入適當的字詞

❶ 如果你想把那些舊雜誌作資源回收，它們應放在那邊的黃色垃圾桶裡。
If you want to recycle those old magazines, they _____ in that yellow bin over there.

❷ 貝蒂酒駕被逮後感到後悔不已。
Betty really regretted it after she was caught driving _____.

Ans ❶ belong　❷ drunk

旅遊英語輕鬆聊　參考答案請掃描使用說明頁上 QR code 下載

If you saw someone making a scene on a plane, would you step in? Why or why not?

71

Unit 18 In-Flight Service
先生，您還需要什麼呢？

實境對話 GO！

🅵 Flight attendant（空服員）　🅴 Edward（艾德華）

Edward is flying economy class. He **presses** the button to call a flight attendant.

🅵 Hello, sir. What can I do for you?

🅴 Do you **serve vegetarian** meals?

🅵 I'm sorry, sir. Vegetarian meals need to be **requested** at least a day before your flight.

🅴 I see. Could I have something to drink?

🅵 **Certainly**. We have coffee, tea, and juice. Which would you like?

🅴 A cup of coffee would be great.

🅵 One moment.

(The flight attendant moves down the aisle. A moment later, she returns with a cup of coffee.)

🅵 Here you are, sir. Is there anything else I can get you?

🅴 Actually, could I **bother** you for another pair of headphones? The ones I have aren't working.

🅵 I'm sorry for the **inconvenience**. I'll be right back with those.

(The flight attendant leaves and returns with a new set of headphones.)

🅵 There you go. I hope these will work all right.

E Thanks. I'm looking forward to watching a film on the in-flight entertainment system. Wait, it seems that the system is not **responding**.

F Really? Let me **assist** you.

E Thanks a lot!

艾德華正搭機坐經濟艙。他按下按鈕來呼叫空服員。

F 先生，您好。有什麼能為您服務的嗎？

E 你們有沒有供應素食餐點？

F 先生，很抱歉。素食餐點需要在您航班出發至少一天前提出要求。

E 了解。那可不可以給我一杯飲料？

F 當然沒問題。我們有咖啡、茶和果汁。您想喝哪個呢？

E 一杯咖啡好了。

F 請稍候。

（空服員沿著走道往後走。過一會兒後，她端了一杯咖啡回來。）

F 先生，這是您要的咖啡。還需要幫您拿些什麼東西嗎？

E 事實上，可否麻煩妳給我另一副耳機？我的這副不能用。

F 很抱歉造成您的不便。我再去拿，很快就回來。

（空服員離開，隨後帶了一副新耳機回來。）

F 這是您要的耳機。希望這副沒問題。

E 謝謝。我很期待用機上娛樂系統看電影。等等，系統好像沒反應。

F 真的嗎？我來協助您。

E 太感謝妳了！

旅遊字詞補給包

① **press** [prɛs] *vt.* 按，壓
Max pressed the switch, but strangely the lamp wasn't working.
麥克斯按下開關，但很奇怪檯燈沒亮。

❷ **serve** [sɝv] *vt.* 供應；端上 (飯菜等)
The hotel staff said breakfast would be served until 9:00 a.m.
飯店人員說早餐會供應到上午九點。

❸ **vegetarian** [ˌvɛdʒəˈtɛrɪən] *a.* 素食的 & *n.* 素食者
The restaurant offers a wide variety of delicious vegetarian dishes.
這間餐廳提供各式各樣美味的素食料理。

❹ **request** [rɪˈkwɛst] *vt.* (正式) 請求，要求
Josephine requested a raise from her boss.
約瑟芬向她的老闆要求加薪。

❺ **certainly** [ˈsɝtn̩lɪ] *adv.* 當然，確實
A: This math question is quite difficult.
B: Yes, it's certainly not easy.
甲：這道數學題目蠻難的。
乙：對啊，它確實不簡單。

❻ **bother** [ˈbɑðɚ] *vt.* 麻煩；打擾
Don't bother driving me home. I'll take a taxi.
不用麻煩你開車送我回家。我會搭計程車。
Kevin kept bothering his sister when she was doing her homework.
凱文在他姊姊寫功課時一直打擾她。

❼ **inconvenience** [ˌɪnkənˈvinjəns] *n.* 不便
The strike is likely to cause inconvenience for the public.
這場罷工可能會造成公眾的不便。
strike [straɪk] *n.* 罷工

❽ **respond** [rɪˈspɑnd] *vi.* 做出反應；回應
Every morning, Doris responds to my greetings with a big smile.
每天早上，桃樂絲都會以燦爛的笑容來回應我的問候。

❾ **assist** [əˈsɪst] *vt.* 協助，幫助
The manager assisted his staff with their tasks.
那位經理協助他的員工完成任務。

經典旅遊實用句

It seems that the [設備] is not responding. ……好像沒反應。

用途：回報設備故障。

系統好像沒反應。
It seems that the <u>system</u> is not responding.
<u>screen</u>　螢幕
<u>entertainment system</u>　娛樂系統
<u>light</u>　燈光

旅人私房筆記

airline meal　飛機餐

飛機餐除了一般餐點之外，還有專屬為乘客不同的需求而特別訂製的餐點，通常需要提前預約喔！

- **Vegetarian Meals**　素食餐
- **Medical Meals**　病理餐
- **Religious Meals**　宗教餐
- **Child Meals**　兒童餐

Try It Out! 試試看

請依句意在空格內填入適當的字詞

① 遊客被要求不得在這間博物館拍照。
Visitors are ＿＿＿＿ not to take photos in this museum.

② 我們用完湯品和沙拉不久後，服務生就端來了主菜。
The waiter ＿＿＿＿ us the main dish shortly after we finished our soup and salad.

Ans ① requested　② served

旅遊英語輕鬆聊

參考答案請掃描使用說明頁上 QR code 下載

If you forgot to order a vegetarian meal, what would you do on the plane?

Unit 19 Asking a Flight Attendant for Help
向空姐求助

實境對話 GO！

K Kevin（凱文）　　**F** Flight attendant（空服員）

A few hours into the flight, Kevin's entertainment center freezes.

K I'm sorry to bother you, but could you help me out?

F Of course, sir. What seems to be the problem?

K My screen froze halfway through *Raiders of the Lost Ark*.

F Let me try restarting the whole system for you. This might take a few minutes, so can I get you anything while you wait?

K I'll take some apple juice and another cheese plate, please.

(The flight attendant goes to restart Kevin's entertainment system.)

F Did that work?

K I'm afraid not. Now, it's frozen at the menu screen.

F I'm very sorry about this. I don't think I'll be able to fix it while we're in the air.

K It's not really your fault, but now it's going to be a long flight.

F What I can do is give you some vouchers for a meal at our VIP area on your flight back.

K That would be great. I guess I'll just have to go back to reading until we get there.

F I feel bad that it's not working, so if you need anything at all, please let me know **right away**.

K I will. Thanks!

> 經過了幾個小時的飛行，凱文的娛樂設施定格當機了。

K 不好意思打擾一下，妳能幫我個忙嗎？

F 當然可以，先生。有什麼問題呢？

K 我《法櫃奇兵》看到一半的時候，螢幕就定格了。

F 讓我試試為您重新啟動整個系統。這可能需要花幾分鐘，所以在您等待的時候，需要我拿點什麼東西給您嗎？

K 請給我蘋果汁，還有乳酪綜合拼盤吧。

（空服員重新啟動凱文的娛樂系統。）

F 有用嗎？

K 恐怕沒有。現在它在選單上定格了。

F 對此我很抱歉。我不認為我能在旅途中把它修好。

K 這又不是妳的錯，但這趟航程恐怕將變得很漫長。

F 我可以做的就是給您一些在回航班機貴賓區用餐的餐券。

K 那太好了。我想我現在只好再來看書，直到我們抵達目的地。

F 我真的很抱歉它不能使用，所以只要您需要什麼，請馬上讓我知道。

K 我會的。謝謝！

旅遊字詞補給包

1. **entertainment** [ˌɛntɚˈtɛnmənt] *n.* 娛樂，樂趣
   The movie was great entertainment for the whole family.
   這部電影是相當適合全家人看的娛樂片。

❷ **freeze** [friz] *vi. & vt.* 當機，定格；(使) 凍結；結冰
All of a sudden, the movie on the screen froze.
螢幕上的電影畫面突然定格了。
It is the nature of water to freeze at zero degrees Celsius.
水的特性就是在攝氏零度時會結冰。

❸ **help sb out**　　幫助某人擺脫困境
Sally always helps people out when they are in trouble.
當別人有困難時，莎莉總會伸出援手。

❹ **halfway** [ˌhæfˈwe] *adv.* 在中途
Eric fell asleep halfway on the bus trip to the next destination.
艾瑞克在坐公車到下個景點的半路上睡著了。

❺ **restart** [riˈstɑrt] *vt.* 重新啟動
After installing the latest updates, I had to restart my computer for them to work.
在電腦上安裝最新的更新後，我勢必要重新啟動電腦讓它們運作。

❻ **right away**　　立刻，馬上
Don't worry. I'll take care of it right away.
別擔心，我會立刻處理。

經典旅遊實用句

I'll take some [飲料] and another [餐點].
我要……和再來一份……。

用途：點餐 / 請求補充餐點。

請給我蘋果汁，還有乳酪綜合拼盤。
I'll take some apple juice and another cheese plate, please.
　　　　　　　　orange juice　柳橙汁　sandwich　三明治

旅人私房筆記 — 機上娛樂相關用語

- **touchscreen**
  [ˈtʌtʃˌskrin]
  觸控螢幕

- **subtitle**
  [ˈsʌbˌtaɪt!]
  字幕

- **volume**
  [ˈvɑljəm]
  音量

- **channel**
  [ˈtʃænl̩]
  頻道 / 節目

- **headphones**
  [ˈhɛdˌfonz]
  （頭戴式）耳機

CH 02 Taking a Flight・搭機

Try It Out! 試試看
請依句意在空格內填入適當的字詞

請立刻到我的辦公室來。
Please come to my office _____ _____ .

Ans right away

旅遊英語輕鬆聊
參考答案請掃描使用說明頁上 QR code 下載

Your screen is broken and your seatmate's is working fine. They're watching the same movie you were. Would you secretly peek at their screen, or just take a nap instead?

79

Unit 20 Last-Minute Gifts
最後才想到買禮物

實境對話 GO！

K Kevin（凱文）　　**F** Flight attendant（空服員）

Kevin realized that he had forgotten to buy some gifts, so he asks the flight attendant for the onboard duty-free magazine.

K Excuse me, miss. Can I take a look at your duty-free catalog?

F Here you go. Can I help you with anything else?

K Actually, yes. I've never bought anything on a plane before. How does it work?

F If you order during the flight, you can get your items right when you leave the plane. They'll be waiting for you at the gate. Our selections are both tax- and duty-free, so you can save a lot.

K Is it any different than buying something at the duty-free shops in the airport?

F Each shop sells different things, and sometimes prices vary. It depends on what you're looking for.

K I want something for my grandmother.

F To tell you the truth, most things are a bit more expensive on board. But if you need to make a quick transfer, it's very convenient and usually worth it.

K Thanks. I don't have a lot of time when I land, so I think I'll get this **perfume** on page 10. I'll pay with my card.

F Here's your **receipt**. Just give this to the assistant as soon as you exit the plane.

凱文突然想到自己忘了買些禮物，所以他向空服員要機上免稅商品雜誌。

K 小姐，不好意思。我可以看一下你們的免稅商品目錄嗎？

F 目錄給您。還有什麼別的事可以幫您嗎？

K 其實是有的。我從來沒有在飛機上買過任何東西。要如何購買呢？

F 如果您在航行期間訂購，你下飛機時就可以拿到您的商品。它們會擺在下機門口那裡。我們供選擇的物品都是免扣稅的，也不用繳稅，所以你可以省超多的。

K 那和機場的免稅店買東西有什麼不同嗎？

F 每家商店出售的商品都不同，而且有時價格會有落差。這都取決於你在找什麼樣的商品。

K 我想帶點東西給我阿嬤。

F 說實話，機上大部分的東西都稍微貴一些。但如果您需要快速轉機的話，這就非常方便了，而且通常很值得。

K 感謝。著陸後我真的沒有太多的時間，所以我想買第十頁上的這款香水。用信用卡付款。

F 這是您的收據。一下飛機後把這個交給我們的地勤人員就可以了。

旅遊字詞補給包

❶ **catalog** [ˋkætəlɔg] *n.* 商品目錄
Nick looked through all the items in the catalog.
尼克瀏覽了商品目錄上的所有品項。

❷ **selection** [səˋlɛkʃən] *n.* 供選擇（選購）的物品
The supermarket has a wide selection of fresh produce.
這間超市有各式各樣的新鮮農產品。

❸ **vary** [ˈvɛrɪ] *vi.* 變化，不同
The weather here varies between cool and very hot.
這裡的天氣變化很大，忽而涼爽忽而酷熱。

❹ **on board**　　在船/火車/飛機上
Please find your seat quickly once you are on board the plane.
登機後請迅速找好您的座位。

❺ **perfume** [pɝˈfjum] *n.* 香水
Linda sprayed a little perfume on her wrists.
琳達在手腕上噴了一點香水。

❻ **receipt** [rɪˈsit] *n.* 收據，發票
Please keep your receipt in case you need to return the item.
請保留您的收據，一旦需要退貨時會用到。

經典旅遊實用句

Can I take a look at your...?　　我可以看一下你們的……嗎？

用途：詢問是否能看免稅商品目錄。

我可以看一下你們的免稅商品目錄嗎？
Can I take a look at your duty-free catalog?
　　　　　　　　　　in-flight shopping guide　　機上購物指南
　　　　　　　　　　magazine　　（機上）雜誌（可指內含免稅商品資訊的雜誌）

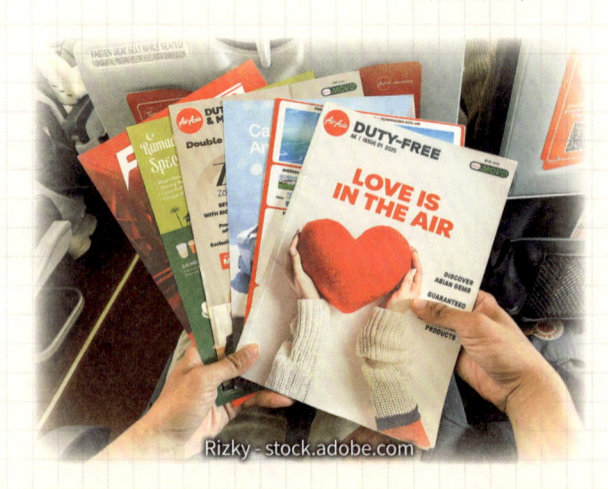

Rizky - stock.adobe.com

旅人私房筆記

免稅商品必買好物

機上購物的商品目錄，我們一起來認識有哪些熱門的免稅商品：

★ Top Picks from Duty-Free

Cosmetics [kɑzˋmɛtɪks]
化妝品

Skincare products
護膚品

Accessories [ækˋsɛsərɪz]
配件

Alcohol [ˋælkəˌhɔl]
酒類

Electronics [ɪlɛkˋtrɑnɪks]
電子產品

Luxury items [ˋlʌkʃərɪ]
奢侈品

CH 02 Taking a Flight・搭機

Try It Out! 試試看　請依句意選出適當的字詞

The store sent us a new ＿＿＿＿＿ with all of their items for next season.
Ⓐ attendant　Ⓑ catalog　Ⓒ receipt　Ⓓ gate

Ans Ⓑ

旅遊英語輕鬆聊　參考答案請掃描使用說明頁上 QR code 下載

Have you ever realized during a trip that you forgot to buy a gift for someone? What did you do?

83

Unit 21 Getting Directions through the Airport
機場裡找路

實境對話 GO！

K Kevin (凱文) **F** Flight attendant (空服員) **R** Receptionist (櫃檯人員)

Kevin's plane has just touched down at Narita International Airport.

K Excuse me, ma'am. Do you know where I have to go to make my connecting flight?

F You're flying to Kaohsiung, right? It looks like gate 33. Head to the left to go through security; then, you can take the tram to gates 20-49.

(Kevin is a little lost after clearing security, so he heads to the information counter.)

R Hello sir, how may I help you?

K I'm trying to get to gate 33, but I can't find the tram.

R OK. You'll have to go downstairs to get to the tram level. Do you see the stairs there?

K Yes.

R Go down those, take a left, and you should be able to follow the signs.

K OK, got it. Is there anywhere to get a quick meal there?

R There's a food court **immediately** downstairs, **as well as** one by gate 30, so you can take your **pick**.

K Great, thanks. My flight starts boarding in 30 minutes. Will that give me enough time?

R Yes. The tram comes every five minutes, and it only takes two minutes to get to the next gate area.

凱文的班機才剛降落至成田國際機場。

K 不好意思，小姐。請問我該去哪裡轉機？

F 您要飛往高雄，是嗎？看來是第三十三號登機門。請往左走過安檢，然後，您可以搭電車到二十至四十九號登機門。

（凱文過了安檢後有點迷路，所以他走向服務檯。）

R 先生，您好，有什麼我可以為您服務的嗎？

K 我想要到三十三號登機門，但我找不到電車。

R 好的。您必須下樓才能到達電車的樓層。您有看到那邊的階梯嗎？

K 有的。

R 從那邊走下去，左轉，接著就可以照著指標走了。

K 好的，我明白了。那裡有地方吃個快餐嗎？

R 一下樓就有一個美食區，三十號登機門旁也有一個，您可以自己選擇。

K 太好了，謝謝。我的班機再三十分鐘就要開始登機了。我會有足夠的時間嗎？

R 會的。電車每五分鐘一班，只要兩分鐘便可到達下一個登機門區。

旅遊字詞補給包

❶ **tram** [træm] *n.* 有軌電車
The tram tracks run along the main street.
電車軌道沿著主要街道行駛。

❷ **downstairs** [ˌdaʊnˈstɛrz] *adv.* 往樓下；在樓下 & *a.* 樓下的
Let's go downstairs for dinner.
我們下樓吃晚餐吧。

❸ **immediately** [ɪˈmidɪɪtlɪ] *adv.* 即刻，馬上
The excited audience clapped their hands immediately after the circus show ended.
興奮的觀眾在馬戲團表演結束後馬上鼓掌。

❹ **as well as**　　也；還有
We enjoyed the beautiful scenery as well as the delicious food.
我們欣賞了美麗的風景，也享受了美味的食物。

❺ **pick** [pɪk] *n.* 選擇 & *vt.* 挑選
The blue dress was Amy's pick for the party.
藍色洋裝是艾咪參加派對的選擇。

經典旅遊實用句

Do you know where I have to go to make my [轉機的班機]?

用途：詢問轉機動線。

請問我該去哪裡轉機？

Do you know where I have to go to make my connecting flight?
　　　　　　　　　　　　　　　　　　　　next flight
　　　　　　　　　　　　　　　　　　　　下一班飛機

旅人私房筆記

在日本機場吃什麼？

日本機場的 food court，總是有琳瑯滿目的日式料理，來為大家介紹幾項日本的代表性美食以及其英文要怎麼說。

ramen 拉麵

sushi 壽司

tempura 天婦羅

udon 烏龍麵

onigiri 飯糰

curry rice 咖哩飯

Try It Out! 試試看　請依句意選出適當的字詞

If you see anything unusual, contact the police ＿＿＿＿＿.
A lastly　　**B** eventually　　**C** slowly　　**D** immediately

Ans　**D**

旅遊英語輕鬆聊　參考答案請掃描使用說明頁上 QR code 下載

Have you ever gotten lost in a big place like an airport or station? How did you feel?

Unit 22 Customs Declaration
海關申報

實境對話 GO！

D Daisy（黛西）　　**O** Officer（海關人員）

Daisy is going through customs at an airport.

D Here is my passport.

O Do you have anything to declare?

D Yes, an apple and a smartwatch a friend gave me.

O I'm sorry, but no fruits can be taken across the border.

D Come on. An apple won't hurt anyone.

O I can't let you take it. An apple may carry an infectious virus or pests.

D Does that mean it may get other people sick?

O No. If it has a disease, there is a possibility that it may infect other crops grown here.

D How about the pests? Do you mean there may be worms inside the apple?

O Exactly. Fresh fruits can carry foreign insects. They can go on to breed and damage our environment and local crops. We'll have to destroy the apple here.

D What a pity! I should have finished it when I was on the airplane.

O Please fill out the Customs Declaration Form and hand it in to that customs agent.

D How much duty do I have to pay for the watch?

O We will determine how much by assessing the duty at the current rate.

D OK. Thank you.

O You're welcome. Next time, don't travel with any produce.

黛西正在機場過海關。

D 這是我的護照。

O 妳有任何東西要申報嗎？

D 有，一顆蘋果和一支朋友送我的智慧型手錶。

O 抱歉，但水果不能帶入境。

D 別這樣。一顆蘋果又不會害到任何人。

O 我不能讓妳帶著它。蘋果可能帶有傳染性的病毒或害蟲。

D 意思是它可能會害其他人生病嗎？

O 不是的。如果它帶有疾病，可能會傳染給其他生長在本地的農作物。

D 那害蟲呢？你的意思是這顆蘋果裡可能有寄生蟲嗎？

O 沒錯。新鮮水果可能帶有外來的昆蟲。牠們會接著繁殖，進而損害我們的環境和當地農作物。我們必須在這裡就把蘋果銷毀。

D 太可惜了！我應該在飛機上就把它吃完的。

O 請填寫這張海關申報單並交給那位報關員。

D 我需要為這支錶付多少關稅？

O 我們會用當前的匯率來估算關稅，以此決定金額。

D 好的。謝謝你。

O 不客氣。下次旅行時別帶著任何農產品。

旅遊字詞補給包

❶ customs [ˈkʌstəmz] *n.* 海關（恆用複數，也可寫為 Customs）
After passing through customs, we could finally leave the airport.
通過海關後，我們終於可以離開機場了。

❷ declaration [ˌdɛkləˈreʃən] *n.* (納稅品等的) 申報
declare [dɪˈklɛr] *vt.* 申報（納稅品等）
Upon arrival, you will need to fill out a customs declaration form.
抵達時，你需要填寫一份海關申報單。
Please declare if you have any food.　如果你有任何食物，請申報。

❸ border [ˈbɔrdɚ] *n.* 國境，邊界
The train crosses the border between France and Spain.
火車穿過法國和西班牙之間的邊界。

❹ infectious [ɪnˈfɛkʃəs] *a.* 傳染性的
infect [ɪnˈfɛkt] *vt.* 感染；傳染
The common cold is highly infectious, especially in crowded places.
普通感冒具有高度傳染性，尤其是在擁擠的地方。
The virus can easily infect people through close contact.
這種病毒很容易透過密集接觸傳染給人。

❺ virus [ˈvaɪrəs] *n.* 病毒
While traveling in South Africa, Anna caught a virus and died.
安娜在南非旅遊時感染到病毒而身亡。

❻ pest [pɛst] *n.* 害蟲
The farmer used insecticide to get rid of the pests in his field.
農夫使用殺蟲劑來去除他田裡的害蟲。

❼ duty [ˈd(j)utɪ] *n.* 關稅
Customs duties are paid on imported products.　進口產品要付關稅。

❽ assess [əˈsɛs] *vt.* 評估；估價
The damage from the storm was difficult to assess.
暴風雨造成的損害難以評估。

❾ rate [ret] *n.* 價格；率，比率
The interest rate will be increased by 0.5%.　利率將調升 0.5%。

經典旅遊實用句

I'm sorry, but no [物品] can be taken across the border.
抱歉，……不能帶入境。

用途：說明禁止入境的物品。

抱歉，但水果不能帶入境。
I'm sorry, but no fruits can be taken across the border.
　　　　　　　　raw meat　　生肉
　　　　　　　　seeds　　　　種子

旅人私房筆記　　過海關常見問題集

A What's the purpose of your visit?　你來本國的目的是？
B I'm here to visit a friend.　我來拜訪朋友。
A Where will you be staying?　你會住哪裡？
B At a hostel.　住青年旅館。
A How long will you be staying in the country?　你計劃在本國待多久？
B For a week.　一週。
A How much currency are you carrying?　你帶了多少貨幣？
B I have US$1,000 cash on me.　我帶了一千美元現金。

Try It Out! 試試看　請依句意在空格內填入適當的字詞

所有的柳橙樹都感染了某種病毒。
All of the orange trees are infected with a _____.

Ans virus

旅遊英語輕鬆聊　參考答案請掃描使用說明頁上 QR code 下載

Do you think airport rules related to bringing food into another country are too strict, or are they necessary?

CH 02　Taking a Flight・搭機

91

Unit 23 Missing Luggage
消失的行李

實境對話 GO!

E Evan（伊凡）　　**G** Ground Staff Member（地勤人員）

Evan had difficulty finding his luggage at the airport, so he asked a ground staff member for help.

E Excuse me, I just got off the plane, and I can't find my luggage on the carousel. I think it's missing.

G I'm sorry to hear that. Let me check the system for you. Can you please provide your flight details and a description of your luggage?

E Sure, my flight was VE123 from San Francisco, and I was connecting from New York. My luggage is black with a red ribbon tied around the handle.

G Thank you for providing the details. Also, could I have your home and email addresses?

E Sure, here you go.

G Let me check the result. It appears that your luggage never made it on the plane in San Francisco.

E How could that happen?

G Sometimes accidents happen and bags aren't sent. But don't worry. Once it's checked, it will **eventually** arrive at its destination.

E What now? Will I have to come back to the airport when it finally arrives?

G No need. We'll send your luggage directly to the address you provided once we find it. You'll also receive emails **informing** you **of** your luggage's **status**. It should arrive at your home within 48 hours.

E That's a relief. Thank you for all your help.

伊凡在機場找不到他的行李，因此他請一位地勤人員幫忙。

E 不好意思，我剛下飛機，而我在行李轉盤上找不到我的行李。我想它不見了。

G 我很遺憾聽到這件事情。讓我為您在系統上確認一下。能請您提供您的班機資訊以及行李描述嗎？

E 好的，我搭的是從舊金山出發的 VE123 班機，我是從紐約轉機的。我的行李箱是黑色的，把手上繫著一條紅絲帶。

G 感謝您提供這些細節。另外，能麻煩給我您的住家地址及電子信箱嗎？

E 當然，給妳。

G 讓我確認一下結果。看來您的行李沒有在舊金山登上飛機。

E 怎麼會發生這種事？

G 有時難免會出差錯，行李沒有被送上飛機。但別擔心。一旦它託運了，最終還是會送達目的地。

E 現在要怎麼做？我需要在它最終抵達時回來機場嗎？

G 不用。一旦我們找到它，我們會直接將您的行李寄到您所提供的地址。您也會收到告知您行李狀態的電子郵件。它應該會在四十八小時內抵達您的住家。

E 那真是令人鬆了口氣。謝謝妳的幫忙。

旅遊字詞補給包

❶ get off... 下（公車、飛機等大型交通工具）
Lily forgot to take her umbrella when getting off the bus.
莉莉下公車時忘了帶走她的傘。

❷ carousel [ˌkærəˈsɛl] *n.* (機場的) 行李轉盤
The baggage carousel at the airport was crowded with passengers waiting for their luggage.
機場的行李轉盤旁擠滿等待行李的旅客。

❸ description [dɪˈskrɪpʃən] *n.* 描述
The police officer asked me to provide a description of the thief's appearance.
警察要我提供對該小偷外貌的描述。

❹ connect from... 從……轉機
Passengers connecting from Tokyo almost didn't get on their next flight in time.
從東京轉機的乘客差點沒能及時趕上下一班班機。

❺ ribbon [ˈrɪbən] *n.* 緞帶，絲帶
Daisy wrapped the gift with a red ribbon.
黛西用一條紅緞帶包裝禮物。

❻ eventually [ɪˈvɛntʃʊəlɪ] *adv.* 最終
Eventually, Sally and Emily became good friends.
最終，莎莉和艾蜜莉成為了好友。

❼ inform sb of / about sth 告知某人某事
Tom informed me about his success in passing the test.
湯姆告訴我他成功通過考試了。

❽ status [ˈstetəs / ˈstætəs] *n.* 狀況，狀態
I haven't received my package yet, so I checked the status of the delivery online.
我還沒有收到包裹，所以我在網上查看了送貨狀態。

經典旅遊實用句

I can't find my [行李] on the carousel.
我在行李轉盤上找不到我的……。

用途：表示行李沒出現在行李轉盤上。

我在行李轉盤上找不到我的行李。
I can't find my luggage on the carousel.
　　　　　　　　 bag　　　包包
　　　　　　　　 suitcase　行李

旅人私房筆記　　　　幾道小撇步，預防弄丟行李

Use luggage tag
使用行李吊牌

Add a unique mark
加上醒目標記

Take a photo of your luggage
拍下行李照片

Keep your baggage claim tag
保留行李條

Try It Out! 試試看　請將下列字詞組合成一正確通順的句子

Frank felt very sad when _____
(of / informed / the police / him) the bad news.

　　　　　　　　　　　　　　　Ans　the police informed him of

旅遊英語輕鬆聊　參考答案請掃描使用說明頁上 QR code 下載

If your luggage got delayed and you were already at your travel destination, what essentials would you need to buy or prepare to get through the next two days?

CH 02 Taking a Flight・搭機

95

Notes

Chapter 03 Accommodations
住宿

Unit 24
Checking In at a Hotel
飯店入住登記

Unit 25
Reservation Trouble
入住搞烏龍

Unit 26
Room Service
客房服務

Unit 27
Complaining to Hotel Staff
天有不「廁」風雲！

Unit 28
Cleaning the Room
打掃客房

Unit 29
Checking Out of a Hotel
從飯店退房

Unit 30
Taking the Hotel Shuttle
搭乘飯店接駁車

97

Unit 24 Checking In at a Hotel
飯店入住登記

實境對話 GO !

R Receptionist（接待人員）　　**B** Beatrice（碧翠絲）

*Beatrice arrives at a hotel and **approaches** the check-in desk.*

R Good afternoon, miss. Could I have your name and see your reservation number?

B Of course. My name is Beatrice Parker. Here's the receipt for my booking.

(Beatrice hands the receptionist a piece of paper.)

R All right, it looks like you have booked the **deluxe suite** for three nights. Is that correct?

B Yes, that's right. Here are my passport and the credit card that I used to book the room. But I was actually wondering if I could use a different card to pay for my stay.

R Sure. That's no problem **so long as** your name is on the card.

B Great! Here, please use this one instead.

(The receptionist copies her ID information and charges her credit card.)

R OK, everything **is in order**. Just sign here, and here is the key card for room number 824.

B Thank you very much.

R Enjoy your stay. If you have any questions, don't **hesitate** to let me know.

B Oh, actually, I would like to **exchange** some of my cash **for** the local **currency**. Can I do that here?

R Certainly. I can help you with that.

碧翠絲抵達飯店後，便走向入住登記櫃檯。

R 午安，女士。能否告訴我您的大名，並出示您的訂房號碼？

B 當然沒問題。我叫做碧翠絲‧帕克。這是我的訂房收據。

（碧翠絲遞了一張紙給接待員。）

R 好的，看來您預訂了三晚的豪華套房。請問正確嗎？

B 對，沒錯。這是我的護照和我用來訂房的信用卡。但我其實想請問我能不能用不同卡片來支付住房費用。

R 當然可以。只要信用卡上是您的大名就沒問題。

B 太好了！這裡，請改刷這張卡片。

（接待人員影印了她的身分資料，並刷了她的信用卡。）

R 好的，一切都準備妥當了。請在這裡簽名就好，然後這是 824 號房的房卡。

B 非常感謝。

R 祝您入住愉快。如果您有任何疑問，請不吝告知。

B 喔，其實我想把身上的一些現金換成本地貨幣。我可以在這裡兌換嗎？

R 沒問題。我可以為您效勞。

旅遊字詞補給包

❶ approach [əˋprotʃ] *vt. & vi.* 接近，靠近
With the big test quickly approaching, it's a good idea to review your class notes. 隨著大考迅速逼近，複習課堂筆記是個好主意。

❷ deluxe [dɪˋlʌks] *a.* 豪華的；高級的
We upgraded to a deluxe cabin on the train for a more comfortable journey. 為了更舒適的旅程，我們升等了火車上的豪華車廂。

❸ suite [swit] *n.* (飯店的) 套房
The bride and groom stayed in the honeymoon suite.
新娘和新郎住在蜜月套房。

❹ **so / as long as...**　　只要……
You will get better as long as you take the medicine the doctor gave you.　只要你服用醫生開給你的藥，你的病就會好轉。

❺ **be in order**　　妥當，無誤
Please check the entire form to make sure all the information is in order.　請檢查整份表格，以確認所有資料都填妥無誤。

❻ **hesitate** [ˈhɛzəˌtet] *vi.* 猶豫，遲疑
Amanda is still hesitating about whether to get married or not.
亞曼達還在猶豫是否要踏上婚姻這條路。

❼ **exchange A for B**　　用 A 交換 B
exchange [ɪksˈtʃendʒ] *vt.* 交換；互換
Karen exchanged the sweater her mom bought her for a larger size.
凱倫將她媽媽買給她的毛衣換成了大一點的尺寸。

❽ **currency** [ˈkɝənsɪ] *n.* 貨幣
Tourists need to exchange their home currency for the local currency.
觀光客需要將他們的本國貨幣兌換成當地貨幣。

經典旅遊實用句

You have booked the [房型] for [天數].
您預定了 [天數] 的 [房型]。

用途：確認房型與住宿天數。

您預訂了三晚的豪華套房。
You have booked the deluxe suite　for three nights.
　　　　　　　　　　　standard room　　two nights
　　　　　　　　　　　標準客房　　　　兩晚
　　　　　　　　　　　twin room　　　　a week
　　　　　　　　　　　雙人房　　　　　一個禮拜

旅人私房筆記 — 飯店房型簡介

single
單人房

double
單床雙人房

twin
雙床雙人房

triple
三人房

suite
套房

presidential suite
總統套房

CH 03
Accommodations・住宿

Try It Out! 試試看
請依句意在空格內填入適當的字詞

午餐時間快到了。我們今天該買什麼呢？
It's _____ lunchtime. What should we get today?

Ans　approaching

旅遊英語輕鬆聊
參考答案請掃描使用說明頁上 QR code 下載

Have you ever stayed at a hotel by yourself? What did you do during check-in?

Unit 25 Reservation Trouble
入住搞烏龍

實境對話 GO！

T Trevor（崔佛）　　**R** Receptionist（接待員）

Trevor goes to the front desk at a hotel to check in for his family.

T Hi, I made a reservation for a double room tonight under the name of Jones.

R OK, let me check our records. Did you make the reservation through our webpage?

T No, actually, our travel agency booked it for us.

R I regret to tell you that they may have failed to make the booking. There's no reservation for Jones this evening.

T What!? No way! I can't believe they made such a serious error.

R I'm afraid so. However, we do have some vacant rooms. Would you like to book one?

T Of course. We just got here from the airport. We want to drop off our things and take a shower.

R I understand. Unfortunately, you might need to wait a little while.

T Why? What's the problem now?

R: Well, the **official** check-out time is 11:00 a.m. After that, housekeeping needs to clean the room. It's 10:30 now, so you'll have to wait until about noon. Is that OK?

T: Fine, we don't have much choice. Please book the room for us.

R: Sure thing. **In the meantime**, you can rest in our lobby or have a bite to eat.

T: Thank you so much.

崔佛走到了飯店櫃檯為家人辦理入住。

T: 妳好，我預訂了一間今晚的雙人房，是用瓊斯這個名字訂的。

R: 好的，我來查看一下訂房記錄。請問您是透過我們的網頁來預訂的嗎？

T: 不是，其實是我們的旅行社幫我們訂的。

R: 很抱歉，他們似乎沒訂成功。今晚的訂房沒有瓊斯這個名字。

T: 什麼！？怎麼可能！我不敢相信他們竟然犯了這麼嚴重的錯誤。

R: 恐怕情況正是如此。不過我們確實還有一些空房。您要訂房嗎？

T: 當然要。我們剛從機場到這邊。我們想把行李放下然後洗個澡。

R: 我了解。不巧的是，你們可能需要稍候一段時間。

T: 為什麼？現在又有什麼問題？

R: 是這樣的，我們表訂的退房時間是上午十一點。在那之後，房務必須進行客房清潔。現在是十點三十分，所以你們必須等到中午左右。請問這樣可以嗎？

T: 好吧，我們也沒有太多選擇。麻煩幫我們訂房吧。

R: 沒問題。在此同時，您可以在大廳稍作休息或是先吃點東西。

T: 非常感謝。

旅遊字詞補給包

❶ **under the name of...**　　用……的名字 / 假名
Samuel Clemens wrote novels under the name of Mark Twain.
塞姆・克萊門斯以馬克・吐溫這個筆名來寫小說。

❷ **regret to V** （正式）對……感到遺憾 / 抱歉
regret [rɪˋgrɛt] *vt.* 後悔 (做了某事)
I regret to inform you that your contract will not be renewed.
很遺憾地通知你，我們不會再跟你續約。
Melody regretted saying those hurtful words to her mom.
美樂蒂後悔說了那些令她媽媽傷心的話。 → hurtful [ˋhɝtfəl] *a.* 傷感情的

❸ **fail to V** 未能 / 無法……
Though James tried hard, he failed to reach the finals this year.
詹姆斯雖然很努力，但他今年未能打入總決賽。

❹ **error** [ˋɛrɚ] *n.* 錯誤 → 也可寫為 typos
The email that Lorenzo wrote contains some typing errors.
羅倫佐寫的電子郵件有一些打字錯誤。

❺ **drop off... / drop...off** （常指汽車）放 / 載…… (至某處)
I want to buy some books, so please drop me off at the store.
我想要買幾本書，所以請讓我在那家店下車。

❻ **official** [əˋfɪʃəl] *a.* 正式的；官方的
The official language of Brazil is Portuguese.
巴西的官方語言是葡萄牙語。

❼ **in the meantime** 在此同時
The next dish will be served soon; in the meantime, you can enjoy your drinks.
下一道菜很快就會上桌；在此同時，你們可以享用飲料。

經典旅遊實用句

We do have some [空房]. Would you like to book one?
我們這邊還有幾間 [空房]。您要訂房嗎？

用途：提供現場可訂房間。
我們確實還有一些空房。您要訂房嗎？
We do have some vacant rooms. Would you like to book one?
　　　　　　　　available rooms　　空房

旅人私房筆記

在大廳背後：讓你舒服入住的無名英雄們！

receptionist / front desk clerk 櫃檯人員

bellboy / bellman / porter 行李員

waiter / waitress 服務生(男) / 服務生(女)

housekeeper / maid 房務人員

doorman 門房

CH 03

Accommodations・住宿

Try It Out! 試試看　請依句意在空格內填入適當的字詞

威爾未能在上個月底前達成銷售目標。
Will _____ _____ meet the sales targets by the end of last month.

Ans failed to

旅遊英語輕鬆聊　參考答案請掃描使用說明頁上 QR code 下載

Have you ever had a problem while checking in at a hotel? What happened and how did you solve it?

105

Unit 26 Room Service
客房服務

Sophia（蘇菲亞） **Receptionist**（接待員）

實境對話 GO！

Sophia calls the front desk of the hotel from her room.

S Hi, I'm Sophia Barns in Room 208. I'm calling because there's a problem with the Wi-Fi.

R Oh, I'm sorry if it's down **at the moment**. We are **updating** our network.

S I see. I was **wondering** when it will be working again.

R It should be back up **within the hour**. Sorry for the inconvenience.

S I understand. Since I'm calling, can I go ahead and order room service for dinner?

R Yes, of course. I can put in your order.

S I'd like a large **pepperoni** pizza and a small order of wings.

R OK, would you care for a drink with your food?

S Yes, please. A bottle of Diet Coke would be great.

R No problem. Your food should arrive in about 30 minutes.

S Thanks. Oh, and one more thing. My **appointment** tomorrow was **rescheduled**, so you can cancel my wake-up call. I'm going to **sleep in**.

R Certainly. That's no problem.

> 蘇菲亞從房間打電話給飯店櫃檯。

S 你好，我是 208 號房的蘇菲亞・巴恩斯。我打來是因為無線網路出了問題。

R 喔，如果目前還無法使用，我要跟您說聲抱歉。我們正在更新我們的網路系統。

S 了解。我想知道網路何時才能恢復正常。

R 應該在一小時之內就會恢復運作。造成不便請您見諒。

S 我明白了。既然我電話都打了，我能順道用客房服務點晚餐嗎？

R 好的，沒問題。我能幫您點餐。

S 我想要一個大的義大利辣臘腸披薩還有一個小份的雞翅。

R 好的，您想要飲料來搭配餐點嗎？

S 好的，麻煩你。請給我一瓶健怡可樂。

R 沒問題。您的餐點將在三十分鐘左右後抵達。

S 謝謝。喔，還有一件事。我明天跟人會面的時間有點更動，所以你可以取消我的電話叫醒服務。我要睡晚一點。

R 當然。沒問題。

旅遊字詞補給包

❶ at the moment　　目前，此刻
Ethan is so busy with work at the moment that he can't discuss the matter with me.
伊森此刻公務繁忙，所以沒空跟我討論這件事。

❷ update [ʌpˋdet] *vt.* 更新
This dictionary is behind the times; it needs to be updated.
這本字典已過時；它需要更新了。

❸ wonder [ˋwʌndɚ] *vt.* 想知道；感到疑惑，納悶
The little girl wondered when her parents would be back from work.
這位小女孩納悶她父母何時會下班回家。

❹ within the hour　　一小時內 (指某事很快就要發生)
Darren told his friends that he would be back from the grocery store within the hour.
達倫告訴他的朋友他一小時內就會從雜貨店回來。

❺ **pepperoni** [ˌpɛpəˈronɪ] *n.* 義大利辣臘腸
I'd like a large pizza with pepperoni and mushrooms.
我要一個大披薩，餡料用義大利辣臘腸加蘑菇。

❻ **appointment** [əˈpɔɪntmənt] *n.* (正式的) 約會，約定
Lawrence left early today because he had an appointment with his dentist.
羅倫斯今天提早離開，因為他跟牙醫有約。

❼ **reschedule** [riˈskɛdʒʊl] *vt.* 重新安排 (行程)
The manager rescheduled her meeting with the team due to some troubles at home.
經理因為家裡有些事情而重新安排與團隊開會的時間。

❽ **sleep in**　　睡到很晚才起床
It is said that sleeping in on weekends is actually not good for your health.
據說週末賴床其實對你的健康有害。

經典旅遊實用句

Hi, I'm Sophia Barns in Room 208. I'm calling because there's a problem with [設備].
你好，我是 208 號房的蘇菲亞‧巴恩斯。我打來是因為⋯⋯出了問題。

用途：打電話報修房內設備。

你好，我是 208 號房的蘇菲亞‧巴恩斯。我打來是因為無線網路出了問題。
Hi, I'm Sophia Barns in Room 208. I'm calling because there's a problem with the Wi-Fi.

　　　　　air conditioner　　空調
　　　　　TV　　電視

旅人私房筆記

Room Services!　教你客房服務的英文怎麼說！

in-room dining / room service 送餐服務	**wake-up call** 叫醒服務	**housekeeping** 房務清潔
laundry service 洗衣服務	**luggage service / bell service** 行李服務	**room maintenance** 客房維修

CH 03　Accommodations・住宿

Try It Out! 試試看　請依句意在空格內填入適當的字詞

琳達此刻無法接電話，因為她正在開會。
Linda can't answer the phone _____ _____ because she's in a meeting.

Ans at the moment

旅遊英語輕鬆聊　參考答案請掃描使用說明頁上 QR code 下載

Have you used room service before? Talk about it.

109

Unit 27 Complaining to Hotel Staff
天有不「廁」風雲！

實境對話 GO！

L Louisa（露意莎）　　**H** Hotel Clerk（服務人員）

Louisa is reporting a problem in her room to the front desk of the hotel.

L Excuse me. I have a problem in my room. The toilet is blocked and I'm worried it may **overflow**.

H I'm sorry for the inconvenience. What is your room number?

L Room 306.

H OK. I'll report it to our **maintenance crew**.

L Thank you. Do you know when they will be able to fix it?

H I'll discuss the matter with them right away. Someone will be sent to the room **shortly**.

L I'm getting ready to go meet some friends. I hope it will be taken care of by the time I return this evening.

H All right, we'll **look into** the matter. If we can't fix it quickly, we'll **get back to** you and arrange for a room **upgrade**.

L That would be nice. Anyway, I'll be back around 8:00 or 9:00 p.m.

H I promise we will take care of the problem **one way or another**.

> 露意莎正在向飯店櫃檯反映她房間的問題。

L 不好意思。我房間有個問題。馬桶堵住了,我擔心水可能會滿出來。

H 很抱歉造成您的不便。請問您的房號是?

L 306 號房。

H 好的。我會彙報給我們的維修小組。

L 感謝。請問你知道他們何時才能修好它嗎?

H 我會立刻和他們討論這件事。很快就會派人去您的房間。

L 我正準備去見朋友。希望在我晚上回來前問題就能解決。

H 好的,我們會查看這個問題。如果我們無法盡速修好它,我們會向您回覆,然後幫您安排客房升等。

L 那太好了。總之,我大概會在晚上八、九點回來。

H 我保證我們會想辦法處理好問題。

旅遊字詞補給包

❶ **overflow** [ˌovɚˈflo] *vi.* 溢出 → 動詞變化為:overflow, overflew, overflown
The bathtub overflowed because I forgot to turn off the water.
我忘了關水龍頭,所以浴缸的水滿出來了。

❷ **maintenance** [ˈmentənəns] *n.* 維護,保養
The building requires regular maintenance to stay in good condition.
這棟建築需要定期維護以保持屋況良好。

❸ **crew** [kru] *n.* 一組工作人員(集合名詞,不可數)
The airline crew welcomed passengers aboard the plane.
航空公司機組人員歡迎乘客登機。

❹ **shortly** [ˈʃɔrtlɪ] *adv.* 不久,很快
Please wait a moment. The clerk will be back shortly.
請稍等一下。店員很快就會回來。

111

❺ look into...　　查看……，調查……
The landlord looked into the electricity problem and gave us an answer the next day.
房東查看了電力的問題，並且隔天就回覆我們。

❻ get back to sb　　答覆/回報某人
Will you come to my party on Saturday? Please get back to me as soon as possible.
你星期六會來參加我的派對嗎？請儘快給我答覆。

❼ upgrade [ˈʌpgred] *n.* 升級
The engineer has made some upgrades to the computer to make it work better.
工程師替這部電腦做了一些升級，以增強它的效能。

❽ one way or another　　設法，無論如何
One way or another, we have to finish this work before the end of the week.
我們必須設法在本週結束前完成這項工作。

經典旅遊實用句

I'll [通報] our maintenance crew.　　我會……給我們的維修小組。

用途：表示將通報維修團隊。

我會彙報給我們的維修小組。
I'll report it to our maintenance crew.
　　　notify　　　　　　　　　　　　　of it.　通知
　　　contact　　聯絡

112

旅人私房筆記　　　常見的入住飯店問題

許多人入住飯店之後，才發現房間有各式各樣的問題，如果發現問題該如何向櫃檯人員反映呢？以下就來看看這些表達方式吧：

- **I want to complain to your manager. / I have a complaint.**
  我要向你們的經理投訴。/ 我有事要投訴。

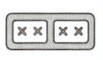

- **None of the power sockets are working.**
  沒有一個電源插座是能用的。

- **The air conditioner isn't working.**
  空調無法運作。

- **There's something wrong with the TV.**
  電視機出了點問題。

- **The water won't drain from the shower.**
  淋浴間無法排水。

- **There is some kind of leak in my room.**
  我的房間有地方漏水。

CH 03　Accommodations・住宿

Try It Out! 試試看
請依句意在空格內填入適當的字詞

① 由於預算緊縮，公司不得不暫緩系統的升級作業。
Due to the tight budget, the company has to put the system _____ on hold.

② 警察正在調查這名兇手的過去。
The police are _____ _____ the killer's past.

Ans ① upgrade　② looking into

旅遊英語輕鬆聊
參考答案請掃描使用說明頁上 QR code 下載

What should you do if something is broken in your hotel room?

113

Unit 28 Cleaning the Room
打掃客房

實境對話 GO！

M▸ Maria（瑪麗亞）　　B▸ Ben（班）

Maria is a housekeeper in a hotel where Ben is currently a guest.

M▸ Housekeeping! May I come in?

B▸ Good morning! Yes, please come in.

M▸ Can I **make up** your room now, sir?

B▸ Yes, please. I'll be heading out for the day in a couple of minutes.

M▸ No problem. Would you like fresh towels or clean sheets today?

B▸ I would, thank you. Oh, there's no toilet paper in the bathroom. Could you **restock** that, too?

M▸ Yes, I'll do that when I **refill** the **toiletries** and take out the trash. Is there anything else that you need? Any extra pillows or a blanket, for example?

B▸ I'm fine for pillows and blankets, but some extra bottles of water would be good.

M▸ I'll take care of it.

B▸ I'd like to skip housekeeping tomorrow, by the way. I'll be having a late night tonight, so I plan to sleep late tomorrow.

M That's not a problem, sir. Please place the "Do Not Disturb" sign on the door when you go to bed tonight so that the rest of my colleagues know not to clean your room.

B I'll do that. Thanks for your help.

M You're welcome, sir. Enjoy the rest of your day.

B Thank you. You, too!

瑪麗亞是飯店房務人員,班目前下榻在該飯店。

M 客房服務!我可以進來嗎?

B 早!請進。

M 先生,我現在可以打掃房間嗎?

B 可以,謝謝。我過幾分鐘就要出去了。

M 好的。您今天需要更換乾淨的毛巾或床單嗎?

B 要,謝謝。喔,浴室裡沒有衛生紙了。可以幫我補嗎?

M 好的,我會在補充盥洗用品和清理垃圾時一併處理。您還需要其他東西嗎?額外的枕頭或毯子之類的?

B 枕頭和毯子不用,但請幫我多放幾瓶水。

M 我會幫您準備。

B 順便跟妳說:明天不需要幫我打掃房間。我今晚會很晚才回來,打算明天睡晚一點。

M 沒問題,先生。今晚睡覺前請把「請勿打擾」的牌子掛在門上,這樣我的同事就知道不需要打掃您的房間。

B 好,謝謝妳的幫忙。

M 不客氣,先生。祝您今天愉快。

B 謝謝,也祝妳愉快!

旅遊字詞補給包

❶ make up 整理;準備
Before the guests arrive, please make up the guest room.
在客人抵達之前,請整理一下客房。

❷ restock [rɪˋstɑk] *vi. & vt.* 再裝滿;為……補貨
We need to restock the office supplies, including paper and pens and other stuff.
我們要補充辦公用品了,像是紙、筆等有的沒的。

❸ refill [ˌriˋfɪl] *vt.* 填補;填充
Could you please refill my coffee cup?
可以幫我續杯咖啡嗎?

❹ toiletries [ˋtɔɪlətrɪz] *n.* 盥洗用品(恆為複數)
The convenience store sells a range of travel-sized toiletries.
這間便利商店販售多種旅行盥洗用品。

❺ disturb [dɪsˋtɝb] *vt.* 打擾;干擾
The loud music next door started to disturb my sleep.
隔壁的吵鬧音樂開始干擾我的睡眠。

❻ colleague [ˋkɑlig] *n.* 同事
Brandon often goes out for lunch with his colleagues from the office.
布蘭登經常和辦公室同事們一起出去吃午餐。

經典旅遊實用句

Can I [打掃房間] now, sir? 先生,我現在可以……嗎?

用途:詢問是否可以現在打掃房間。

先生,我現在可以打掃房間嗎?
Can I make up your room now, sir?
　　　 tidy up the room　　整理房間

旅人私房筆記

Toiletries　盥洗用品

- **shampoo** 洗髮精
- **comb** 梳子
- **conditioner** 潤髮乳
- **toothpaste** 牙膏
- **earplugs** 耳塞
- **shaver** 刮鬍刀
- **soap** 肥皂

Try It Out! 試試看
請依句意在空格內填入適當的字詞

你能在我們親戚來之前把客房整理好嗎？
Could you _____ _____ the guest room before our relatives arrive?

Ans make up

旅遊英語輕鬆聊
參考答案請掃描使用說明頁上 QR code 下載

What hotel item do you always want more of? Why?

CH 03 Accommodations・住宿

117

Unit 29 Checking Out of a Hotel
從飯店退房

實境對話 GO！

C Clerk（接待人員）　　**A** Alice（愛麗絲）

Alice walks up to the front desk of her hotel.

C Good morning, how may I help you?

A I'd like to check out.

C OK, no problem. May I have your name and room number please?

A Sure. My name is Alice Smith, room number 1128.

C All right, Ms. Smith. Please wait a moment while one of our staff checks the room. We need to make sure there is no damage. Otherwise, we'll need to charge a fee.

A Oh, I see. Well, that shouldn't be a problem.

(Three minutes later)

C I'm afraid our staff member found that you left the room in a bad state, so we'll need to charge your card.

A What? How is that possible? I didn't break anything. I left everything exactly how it was. Call again to make sure.

C I see. I'm so sorry. This was my fault. I must have told the staff the wrong room number.

🅐 You almost charged me money for something I didn't do. This is quite **unprofessional**.

🅒 I'm very sorry. I would like to offer you a **discount** on your next stay here, if you should decide to stay again.

🅐 Since you are so sincere, I can **accept** it and forgive you. Please be more careful next time.

愛麗絲走到她飯店的前檯。

🅒 早安,有什麼能協助您的嗎?

🅐 我想要退房。

🅒 好的,沒問題。可以請您給我您的大名及房號嗎?

🅐 當然。我的名字是愛麗絲・史密斯,房號是 1128。

🅒 好的,史密斯女士。請您稍待我們員工檢查房間。我們需要確保房內沒有任何損壞。否則,我們會需要收取一筆費用。

🅐 噢,了解。嗯,這應該不成問題。

(三分鐘後)

🅒 不好意思,我們的員工發現您的房間狀態很糟,所以我們會需要從您的卡片中扣款。

🅐 什麼?這怎麼可能?我沒有破壞任何東西。我將一切恢復原樣。再打一次確認看看。

🅒 我了解了。我很抱歉。這件事是我的錯。我一定是告訴員工錯誤的房號了。

🅐 你幾乎要因為我沒有做的事情扣我錢。這相當不專業。

🅒 我非常抱歉。如果您決定再次入住,我想在您下次入住時提供您一個折扣。

🅐 既然你非常地誠懇,我可以接受並原諒你。請你下次要更加注意。

旅遊字詞補給包

❶ make sure (that) + S + V　確定 / 務必……
Make sure that the door and the windows are locked before you go to sleep.
在你睡覺前要確保門窗有鎖好。

❷ damage [ˋdæmɪdʒ] *n.* 損害，損傷
The powerful storm did a lot of damage to the village.
這場強大的暴風雨對該村莊造成很大的損害。

❸ state [stet] *n.* 狀態；情況
Grace was in a poor state of health.
葛蕾絲的健康狀況不佳。

❹ exactly [ɪgˋzæktlɪ] *adv.* 確切地，精確地
Luke has exactly the same car as you.
路克有一輛跟你完全相同的車。

❺ fault [fɔlt] *n.* 錯誤，過錯
It's not anyone's fault for the loss of the game.
輸了比賽不是任何人的錯。

❻ unprofessional [ˌʌnprəˋfɛʃənḷ] *a.* 不專業的
Wearing casual clothes to a formal business meeting is unprofessional.
穿著休閒服參加正式的商務會議是很不專業的。

❼ discount [ˋdɪskaʊnt] *n.* 折扣
They are offering a 20% discount on all summer clothing.
他們提供所有夏季服裝八折優惠。

❽ accept [əkˋsɛpt] *vt.* 接受
Jerry accepted Lucy's apology and forgave her.
傑瑞接受了露西的道歉並原諒她。

經典旅遊實用句

I'd like to [退房].　　我想要……。

用途：主動表示要退房。

我想要退房。
I'd like to check out.
　　　　　settle my bill　　　結清帳單
　　　　　complete my stay　　辦理退房

旅人私房筆記

Check out 時可能遇到的麻煩情境

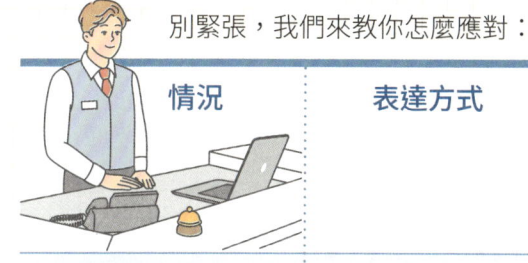

別緊張，我們來教你怎麼應對：

情況	表達方式
櫃檯說你有在迷你吧消費，但你沒有。	I'm sorry. I didn't take anything from the minibar. 抱歉，我沒有拿過迷你吧裡的東西。
房間被誤認為有損壞。	Could you please double-check the room? 你們可以再去房間檢查一遍嗎？
帳單金額錯誤。	There seems to be a mistake on the bill. 帳單好像有錯。

Try It Out! 試試看

請依句意在空格內填入適當的字詞

如果你要主辦派對，你必須確定有足夠的食物和飲料。
If you host a party, you'll need to _____ _____ there's enough food and drink.

Ans make sure

旅遊英語輕鬆聊

參考答案請掃描使用說明頁上 QR code 下載

Have you ever forgotten anything in a hotel room? What was it?

121

Unit 30 Taking the Hotel Shuttle
搭乘飯店接駁車

實境對話 GO!

C Cara（卡拉）　　**J** Jacob（雅各）

Jacob has just landed at the airport. He is speaking to Cara, a receptionist at the Carlton Hotel, on the phone.

C Good morning, Carlton Hotel. Cara speaking. How may I help you?

J Hi. I just landed at the airport, and I was wondering if you provide a shuttle service to the hotel.

C We do **indeed** offer a **complimentary** shuttle service.

J Where can I catch it?

C Which **terminal** are you at, sir?

J Terminal 2.

C Our shuttle picks up guests outside the Terminal 2 Arrivals Hall, near Exit D. If you come out of that exit, the bus stop is just to the right of the taxi **rank**.

J Got it, thanks. Will I need to wait long?

C Providing that the driver does not get held up in traffic, the next shuttle should be arriving in around ten minutes. Look for a black **vehicle** with the logo of the hotel on the side.

J I've got three large **suitcases**—will that be OK?

C That's absolutely fine. There's plenty of space, and the driver will help you with your bags. Just make sure to show him your reservation email to prove you're staying at our hotel.

J Perfect. Thanks for your help.

C No problem. It should take around 20 minutes to reach the hotel. Safe travels!

雅各的飛機剛降落在機場。他正和卡爾頓飯店的櫃檯人員卡拉講電話。

C 早安，這裡是卡爾頓飯店，我是卡拉。請問需要什麼服務？

J 嗨，我的飛機剛到機場，想問一下你們有飯店接駁車服務嗎？

C 我們是有提供免費接駁車服務喔。

J 我要在哪裡搭車？

C 請問您現在在哪個航廈？

J 第二航廈。

C 我們的接駁車會在第二航廈入境大廳外面接駁旅客，在出口 D 附近。您從出口出來後，接駁車站就在計程車招呼站右邊。

J 了解，謝謝。要等很久嗎？

C 如果司機沒有碰上塞車的話，下一班接駁車大概十分鐘後就會到。請找一臺車側有我們飯店標誌的黑色車輛。

J 我有三個大行李箱 —— 這樣可以嗎？

C 當然沒問題。車子的空間很足夠，而且司機會幫您搬行李。請務必出示您的訂房確認信給司機看，證明您是我們飯店的房客。

J 太好了，謝謝妳的幫忙。

C 不客氣。從機場到飯店大約需要二十分鐘。
祝您一路平安！

旅遊字詞補給包

❶ shuttle [ˈʃʌtl̩] *n.* 接駁車
The company runs an employee shuttle to the office park.
這間公司提供員工接駁車前往辦公園區。

❷ indeed [ɪnˈdid] *adv.* 確實，的確
Dolly is indeed a talented artist.
朵莉的確是一位才華洋溢的藝術家。

❸ complimentary [ˌkɑmpləˈmɛntərɪ] *a.* 免費贈送的；讚美的
The hotel offers complimentary breakfast for all guests.
這間飯店為所有住客提供免費早餐。
Harold's performance was met with complimentary reviews.
哈洛德的表演獲得好評。

❹ terminal [ˈtɝmənl̩] *n.* 航廈；總站
We need to go to Terminal 2 for our flight to Tokyo.
我們要去第二航廈搭乘前往東京的班機。

❺ rank [ræŋk] *n.* 計程車候車站
You can usually find a taxi quickly at the taxi rank near the shopping mall.
在購物中心附近的計程車招呼站通常很快就能找到計程車。

❻ vehicle [ˈviɪkl̩] *n.* 車輛；任何有輪子的運載工具
The bicycle is a popular and eco-friendly vehicle in many cities.
自行車在許多城市是一種受歡迎且環保的交通工具。

❼ suitcase [ˈsutˌkes] *n.* (有把手的) 行李箱
The customs officer asked Jerry to open his suitcase.
海關人員要求傑瑞打開他的行李箱。

經典旅遊實用句

Our shuttle... outside the Terminal 2 Arrivals Hall, near Exit D. 我們的接駁車會在第二航廈入境大廳外面……，在出口 D 附近。

用途：說明接駁車等候地點。

我們的接駁車會在第二航廈入境大廳外面接駁旅客，在出口 D 附近。
Our shuttle picks up guests outside the Terminal 2 Arrivals Hall,
　　　　　 stops　　 停靠
near Exit D.

旅人私房筆記

常見的機場標誌

- **Shuttle Bus** 接駁巴士
- **Public Taxi** 計程車
- **Restaurants** 餐廳
- **Car Park** 停車場
- **Customs** 海關
- **Information** 服務臺

Try It Out! 試試看

請依句意在空格內填入適當的字詞

這間商店提供免費禮品包裝服務。
The store offered a _____ gift wrapping service.

Ans complimentary

旅遊英語輕鬆聊

參考答案請掃描使用說明頁上 QR code 下載

What do you usually do while waiting for transportation, like a hotel shuttle or bus?

Notes

Chapter 04 Transportation
交通

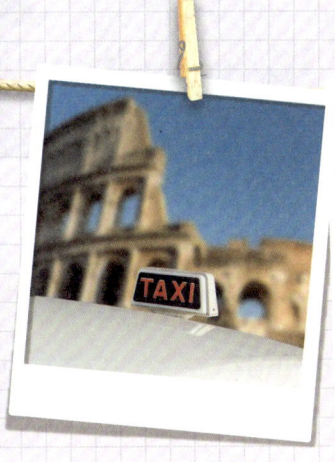

Unit 31
When in Rome
羅馬輕鬆遊

Unit 32
Renting a Car
租輛車子好代步

Unit 33
Filling Up the Tank
加滿油

Unit 34
Buying a Train Ticket
購買火車票

Unit 35
Taking a Taxi
搭計程車

Unit 36
Taking the Bus
搭公車

Unit 37
Getting on the Wrong Bus
公車迷途記

Unit 31 When in Rome
羅馬輕鬆遊

實境對話 GO！

J Jennifer（珍妮佛）　　**B** Basil（貝索）　　**M** Matteo（馬泰奧）

Jennifer and her boyfriend Basil just arrived at the airport in Rome.

J Here we are in Rome. How should we get around?

B We could get a car, but parking will be difficult.

J How about something smaller? Are you thinking what I'm thinking?

B Vespa?

J Yes! They are the greatest scooters ever—true Italian classics. Let's rent one!

(At the rental office.)

M Good morning, my name is Matteo. How may I help you?

B I'd like to rent a Vespa for three days.

M That'll be €50 a day. I'll need your passport and your motorcycle license.

B Oh shoot, I left my motorcycle license at home!

J No worries, honey, I've got mine.

M Thank you, ma'am. I trust that you'll be driving then?

J Of course. What colors do you have?

M We have white, blue, red, green, orange, or pink.

J We'd like the pink one.

B Oh man! You're really trying to punish me here, aren't you?

J Of course. Next time, you won't forget your license. Matteo, where do I sign?

M Right here. All done; here are your keys. Enjoy your time in Rome!

珍妮佛和男友貝索剛抵達羅馬的機場。

J 我們現在到羅馬囉。我們要怎麼去四處逛逛呢？

B 我們可以去租輛車來開，但找停車位是個難題。

J 不如租小一點的交通工具怎麼樣？你跟我在想的是同樣的東西吧？

B 偉士牌機車？

J 沒錯！它們是史上最棒的機車，真正的義大利經典之作。咱們租一部來騎吧！

（在租車辦公室裡。）

M 早安，我叫馬泰奧。有什麼我可以為您效勞的嗎？

B 我想租一部偉士牌機車三天。

M 那每天是五十歐元喔。我需要你的護照及機車駕照。

B 糟糕，我把機車駕照放在家裡了！

J 親愛的，別擔心，我有帶我的。

M 謝謝您，女士。我想待會是您騎車吧？

J 當然。你們的偉士牌有什麼顏色呢？

M 我們有白色、藍色、紅色、綠色、橘色，也有粉紅色的。

J 那我們租粉紅色的那一部。

B 不是吧！這是一種懲罰嗎？

J 當然囉。這樣下次你就不會忘記帶駕照了。馬泰奧，我該在哪裡簽名呢？

M 就在這。手續完成了，你們的鑰匙在這。好好享受你們在羅馬的時光囉！

funkenzauber - stock.adobe.com

CH 04 Transportation・交通

129

旅遊字詞補給包

❶ **get around**　各處旅行；四處走動
The best way to get around the island is to rent a car.
在這個島上趴趴走最好的方式是租車。

❷ **scooter** [ˈskutɚ] *n.* (小型)機車；滑板車
Many delivery drivers ride scooters to deliver food in the city.
許多外送員騎機車在市區送餐。

❸ **classic** [ˈklæsɪk] *n.* 經典；傑作，經典作品
Mario Kart is a classic for video game fans.
《瑪利歐賽車》是電玩迷口中的經典。

❹ **rent** [rɛnt] *vt.* 租借
rental [ˈrɛntl̩] *n.* 租用的東西；出租；租賃
Angela would like to rent a car for just a day.
安琪拉想租一部車，一天即可。

❺ **license** [ˈlaɪsn̩s] *n.* 執照；許可證
The pet shop needs a license to sell animals.
這間寵物店需要許可證才能販售動物。

❻ **punish** [ˈpʌnɪʃ] *vt.* 處罰／懲罰(做錯事或犯法的人)
If Terry is the one that made the mistake, he should be punished.
如果犯錯的人是泰瑞的話，他就應該接受懲罰。

經典旅遊實用句

What [顏色／款式] do you have?　你們有什麼……？

用途：詢問選項(顏色／款式等)。

你們有什麼顏色？
What <u>colors</u> do you have?
　　　<u>models</u>　款式
　　　<u>sizes</u>　尺寸
　　　<u>types</u>　類型
　　　<u>styles</u>　樣式

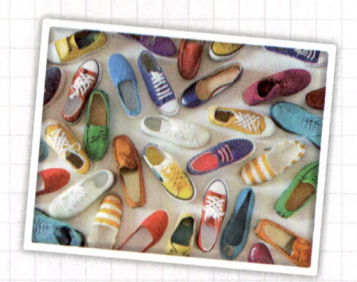

旅人私房筆記　　騎 Vespa 逛羅馬！

我們來看看有哪些著名的景點：

Colosseum 羅馬競技場

Trevi Fountain 特雷維噴泉

Pantheon 萬神殿

Spanish Steps 西班牙階梯

Trastevere 特拉斯提弗列區

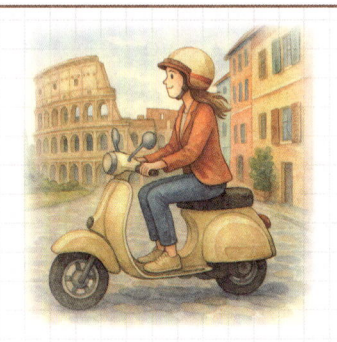

Try It Out! 試試看　　請依句意在空格內填入適當的字詞

若不遵守法律，你將會受罰。
You'll be _____ if you do not obey the law.

Ans punished

旅遊英語輕鬆聊　　參考答案請掃描使用說明頁上 QR code 下載

What color Vespa would you choose, and what does that color say about your personality?

CH 04　Transportation・交通

131

Unit 32 Renting a Car
租輛車子好代步

實境對話 GO！

A Alan（艾倫）　　**C** Clerk（店員）

Alan just arrived in a foreign country and is getting ready to rent a car.

A Hello, I'm Alan Moore and I'm here to rent a car. I reserved one online a few days ago.

C Certainly. Do you have your reservation code? We should have sent you an email.

A Yes, I've got it right here.

C Excellent. Now, I just need to see your passport and your international driver's license.

A Here you go.

C This says that you want to rent a full-size SUV, correct?

A Yes, I'm going to be driving quite a bit, and I'm planning to visit the mountains as well.

C OK. The SUV rental price is US$40 for each day for the first five days. After that, it is US$30 a day for a period of up to 20 days.

A That's fine. I'll need it for about a week.

C We also have to charge a basic insurance fee of US$25.

A That's reasonable, I suppose.

C I'll make the charges to the credit card you provided online. Then, I just need you to sign these forms.

A No problem.

C OK, here are your keys and registration info. The SUV you requested is in the Green Lot, in space 24. You have a safe trip.

A Great. Thank you so much!

艾倫剛抵達某個外國國家，正準備要租一輛車。

A 妳好，我叫艾倫‧摩爾，我是來租車的。我前幾天在網路上預約了一輛。

C 好的。請問您的預約代碼是？本公司應該有寄給您一封電子郵件才對。

A 有，我準備好了，就在這裡。

C 太好了。我現在只需要確認一下您的護照及國際駕照。

A 來，給妳。

C 上頭顯示您想要租一輛大型運動休旅車，對吧？

A 是的，我會開很長一段距離，而且我還打算開到山上去。

C 好。休旅車的租金是前五天每天四十美元。超過五天後，之後多達二十天的每日租金是三十美元。

A 可以接受。我大概需要租一個星期。

C 我們還必須酌收二十五美元的基本保險費。

A 我想這還算合理。

C 那麼我會用您在網路上登記的那張信用卡來刷卡收費。接著，我就只要請您在這些表格上簽名就行了。

A 沒問題。

C 好了，這裡是您的鑰匙和車輛登記資料。您所預約的休旅車就停在綠色停車場的 24 號車格。祝您一路平安。

A 太好了。非常感謝！

旅遊字詞補給包

❶ international [ˌɪntɚˈnæʃənḷ] *a.* 國際的
Jamie learned English by watching the international news.
傑米透過收看國際新聞來學習英文。

❷ up to + 數字　　多達……
Up to 20 people can be on board this hot air balloon.
這顆熱氣球上可乘載多達二十個人。

❸ insurance [ɪnˈʃʊrəns] *n.* 保險
The medical insurance system benefits the elderly and disabled citizens a lot.
健保系統讓年老及殘障市民受惠不少。

> disabled [dɪsˈebḷd]
> *a.* 殘疾的，喪失能力的

❹ registration [ˌrɛdʒɪˈstreʃən] *n.* 車輛登記證，行照；登記，註冊
Student registration tends to start in late August.
學生註冊手續往往都在八月底開始辦理。

❺ request [rɪˈkwɛst] *vt.* 要求，請求
All staff members are requested to wear factory uniforms.
所有職員都被要求穿工廠制服。

經典旅遊實用句

I'm here to rent... I reserved one online a few days ago.
我是來租……的。我前幾天在網路上預約了一輛。

用途：租賃車輛使用。

我是來租車的，我前幾天在網路上預約了一輛。
I'm here to rent a car. I reserved one online a few days ago.

a scooter	機車
a van	廂型車
an SUV	休旅車

旅人私房筆記

常見出租車款有哪些

compact car
小型車

economy car
經濟型車

mid-size / intermediate car
中型車

full-size car
大型車

SUV (sport utility vehicle)
運動休旅車

van
廂型車

Try It Out! 試試看　請依句意在空格內填入適當的字詞

① 在這次特賣會中，多數商品的折扣來到六折之多。
Most of the products have been discounted by ＿＿＿＿＿ ＿＿＿＿＿ 40% in this big sale.

② 警方正在尋找一輛車牌登記為 H06-RBY 的小型灰色轎車。
Police are looking for a small gray car with the ＿＿＿＿＿ number H06-RBY.

Ans ① up to　② registration

旅遊英語輕鬆聊　參考答案請掃描使用說明頁上 QR code 下載

Have you ever rented a car in another country? What was the experience like?

Unit 33 Filling Up the Tank
加滿油

實境對話 GO！

C Carter（卡特）　　**B** Bess（貝絲）

Bess is pulling into a gas station where Carter works as an attendant.

C Good morning. What can I do for you?

B Fill her up, please.

C Certainly, miss. Would you like regular or premium unleaded, or diesel?

B Oh! This is embarrassing… It's a rental, and I don't know what kind of fuel it takes!

C You definitely don't want to put the wrong fuel in! If it's a rental, though, the key tag or rental paperwork might say.

B You're right! There's a label on the key tag—it says unleaded. Regular is fine, thanks. Sorry about that.

C Not a problem. Being near a popular tourist area, we get a lot of rental cars looking to fill up. You want a full tank, right?

B Yes, please.

C *(Inserts the nozzle and fills up the tank.)* Would you like to pay with cash or card?

B Card, please. Do I pay here or inside?

C You can pay here at the pump—I've got the machine. That'll be $52.55.

B Here's my card.

C Thanks. Here are your card and receipt. Is there anything else I can help you with? Would you like your tires or oil checked?

B There's no need, thanks.

C OK, miss, you're good to go. Have a nice day.

B You, too!

貝絲正把車開進卡特工作的加油站。

C 早安，請問要怎麼加？

B 請加滿。

C 好的，小姐。您需要普通無鉛汽油、高級無鉛汽油還是柴油？

B 喔！糟了⋯⋯這臺車是租的，我不知道該加什麼油！

C 千萬不能加錯油喔！不過如果是租車的話，鑰匙吊牌或租車文件上應該會寫。

B 對耶！鑰匙吊牌上有個標籤寫無鉛汽油。幫我加普通的就好，謝謝。不好意思。

C 沒事。我們這裡離熱門觀光區很近，常常有租車客人來加油。您要加滿對吧？

B 對，麻煩你。

C （插入油槍並加滿油。）您要付現還是刷卡？

B 刷卡。在這裡刷還是去裡面？

C 在這裡就可以了，我這裡有機器。總共是 52.55 美元。

B 這是我的卡。

C 謝謝。這是您的卡和收據。還需要其他服務嗎？要不要檢查輪胎或機油？

B 不用了，謝謝。

C 好了，小姐，您可以上路了。祝您有個美好的一天。

B 你也是！

旅遊字詞補給包

❶ **premium** [ˈprimɪəm] *a.* 高級的
The car offers a premium sound system as an optional upgrade.
這輛車可以選配高級音響系統作為升級。

❷ **unleaded** [ʌnˈlɛdɪd] *a.* 無鉛的
The price of unleaded gasoline has been rising recently.
無鉛汽油的價格最近一直在上漲。

❸ **diesel** [ˈdizḷ] *n.* 柴油
Many trucks and buses run on diesel fuel.
許多卡車和巴士使用柴油作為燃料。

❹ **fuel** [ˈfjuəl] *n.* 燃料
The airplane needs a large amount of fuel for its long flight.
這架飛機長途飛行需要大量的燃料。

❺ **tag** [tæg] *n.* 標牌，標籤
The price tag on the shirt said it was on sale.
襯衫上的價格標籤顯示它正在特價。

❻ **nozzle** [ˈnɑzḷ] *n.* 噴嘴
The fuel nozzle needs to be inserted correctly into the car's gas tank.
加油槍的噴嘴需要正確地插入汽車的油箱。

經典旅遊實用句

It's a [某種類型的車], and I don't know what kind of fuel it takes. 這是……，我不知道該加什麼油。

用途：說明你不了解車輛的加油需求。

這臺車是租的，我不知道該加什麼油。
It's a <u>rental</u>, and I don't know what kind of fuel it takes.

   borrowed car 借來的車
   company car 公司車

138

旅人私房筆記

汽車相關檢查與緊急處理實用字詞

flat tire
爆胎

low tire pressure
胎壓不足

engine light on
引擎警示燈亮

jump start
跳接發動（如果電瓶沒電）

Try It Out! 試試看　請依句意在空格內填入適當的字詞

這家航空公司提供豪華經濟艙座位，有更寬敞的腿部空間。
The airline offers _____ economy seats with more legroom.

Ans premium

旅遊英語輕鬆聊　參考答案請掃描使用說明頁上 QR code 下載

How do you handle driving on the "other" side of the road when traveling?

CH 04
Transportation・交通

139

Unit 34 Buying a Train Ticket
購買火車票

實境對話 GO !

C Clerk（售票員）　　**S** Stella（史黛拉）

Stella is trying to buy a ticket at a train station.

C Hello, ma'am. How can I help you today?

S Hello. I would like to book a **one-way** train ticket to Busan for tomorrow morning.

C No problem. There are several trains **leaving for** Busan in the morning.

S How much is a ticket? And how long is the **journey**?

C Well, that **depends on** which train you take. The earliest KTX **Express** will get you there in a little over two hours. The economy-class seat is about 60,000 won.

S Oh, that sounds great. I'll take that one.

C That train will leave at 8:10 a.m.

S Actually, that's quite early. Is there another express train leaving closer to noon?

C Let's see. There is an 11:45 a.m. train. But unfortunately, the economy-class tickets are already fully booked. There is still first-class seating available, though.

S I see. How much is a first-class ticket?

C That will cost about 90,000 won.

S I shouldn't spend that much. Are there any other trains with seats still available?

C Yes, there is a 9:50 a.m. train, but it will take about three hours to arrive in Busan.

S That's fine. I'd like a ticket for the 9:50 a.m. train.

C No problem. That will be 56,000 won.

史黛拉在車站準備買一張火車票。

C 哈囉，小姐。今天有什麼能為您服務的嗎？

S 哈囉。我想訂一張明天早上前往釜山的單程票。

C 沒問題。早上有好幾班前往釜山的列車。

S 一張票要多少錢？還有，車程要多久？

C 嗯，這要看您搭的是哪一種列車。最早一班的韓國高鐵可以讓您在兩個小時又多一點點的時間內抵達。經濟車廂的座位大概要六萬韓圓。

S 噢，那聽起來很棒。我想要搭那班車。

C 列車會在上午八點十分出發。

S 那樣其實太早了。有沒有其他比較接近中午出發的特快車？

C 讓我看看。有一班上午十一點四十五分的列車。可惜的是，經濟車廂的票都被訂光了。不過，還是有頭等車廂的空位。

S 我了解了。一張頭等車廂的票要多少錢？

C 大概要九萬韓圓。

S 我不該花那麼多錢的。還有沒有其他有空位的列車？

C 有的，有一班上午九點五十分的列車，不過要花大約三小時才會到釜山。

S 那沒關係。我要一張上午九點五十分那班車的車票。

C 沒問題。這樣是五萬六千韓圓。

旅遊字詞補給包

❶ **one-way** [ˌwʌnˈwe] *a.* 單程的
   round-trip [ˈraʊndˌtrɪp] *a.* 來回的
   Dave bought a one-way ticket to London because he wasn't sure when he would return.
   戴夫買了一張到倫敦的單程票，因為他不確定何時會回來。
   Eliot prefers to take a round-trip bus because it's more convenient than finding separate transportation back.
   艾略特比較喜歡搭乘來回巴士，因為這比另外找回程交通工具方便。

❷ **leave for +** 地方　　前往某地
   Ransom will leave for America to attend graduate school this autumn.　蘭森今年秋天會前往美國就讀研究所。

❸ **journey** [ˈdʒɝnɪ] *n.* 旅程
   Carlos treasures the memories of his journey across Europe.
   卡洛斯非常珍惜他旅行歐洲的回憶。

❹ **depend on...**　　取決於……
   Whether Lisa can come to the party depends on how much work she has.　麗莎能不能參加派對取決於她的工作量。

❺ **express** [ɪkˈsprɛs] *n.* 快車 & *a.* 快速的
   an express train　　特快車
   We decided to take an express train to Hualien to save time.
   我們決定搭乘一班特快列車去花蓮以節省時間。

經典旅遊實用句

I would like to book a [票種] train ticket to [地方].
我想訂一張前往 [地方] 的 [票種] 車票。

> 用途：說明票種、目的地與出發時間，適合購票起手句。
> 我想訂一張明天前往釜山的單程車票。
> I would like to book a <u>one-way</u> train ticket to <u>Busan</u>.
>
> | | | | |
> |---|---|---|---|
> | round-trip | 來回票 | Kyoto | 京都 |
> | return | 來回票 | Munich | 慕尼黑 |

旅人私房筆記

火車上關於座位的實用句子

除了文中出現的對於車廂的要求外，如果乘客對於位子有所要求或是想要跟別的乘客換座位的話，要怎麼說呢？我們一起來學學這些實用句。

A Would you like a window seat or an aisle seat?
B I would like a window / an aisle seat, please.
　 I'll take a window / an aisle seat.

A 您要靠窗還是靠走道的座位呢？
B 麻煩給我一個靠窗 / 靠走道的位子。
　 我選擇靠窗 / 靠走道的位子。

I wonder if I could switch to an aisle / a window seat.
我想知道我是否可以換到靠走道 / 靠窗的座位。

和別的旅客換座位時又要怎麼說？

Could I change seats with you?
我可以跟你換座位嗎？

Do you mind if we switch seats?
Would you mind if we switched seats?
你介意我們換一下座位嗎？

Try It Out! 試試看　請依句意在空格內填入適當的字詞

這本小說是關於一個英雄穿越魔法國度的旅行。
This novel is about the hero's _____ across a magical world.

Ans journey

旅遊英語輕鬆聊　參考答案請掃描使用說明頁上 QR code 下載

| What's the worst seatmate you can imagine on a train?

CH 04

Transportation・交通

Unit 35 Taking a Taxi
搭計程車

實境對話 GO！

C Caroline（卡洛琳）　　**P** Pedro（佩卓）

Caroline is trying to find a taxi and hails one driven by Pedro.

C Excuse me, are you available?

P Yes, I am—hop in.

C Oh, great! Finally! I've been trying to get a taxi for the last 10 minutes.

P Most people book them via the app these days, so there are fewer cabs available to be hailed on the street.

C I'm a tourist, so I haven't got the app.

P No worries. Where would you like to go?

C I need to go to Central Station, please.

P No problem. Do you want the scenic route or the fastest route?

C I'd better go for the quickest one. My train leaves in 30 minutes.

P Got it. Even in this traffic, the journey should only take 15 minutes.

C Great, thank you. How much is it to the station?

P It should be around 12 euros, but the meter will tell us exactly once we arrive.

C Can you drop me at the west entrance, please? My train leaves from **Platform** 5.

P Sure thing.

C And can I pay by card?

P Cash or card is fine. I have a card machine here for you to use.

C Thank you.

P You're welcome. Please put your seat belt on, sit back, and enjoy the ride.

卡洛琳在攔計程車，攔到了佩卓開的車。

C 不好意思，可以載我嗎？

P 可以，請上車。

C 太好了！終於叫到車！我已經等了十分鐘都叫不到車。

P 現在大多數人都是透過 app 叫車，所以路邊能攔到的車比較少。

C 我是觀光客，沒有那個 app。

P 沒關係，妳想去哪裡？

C 我要去中央車站，謝謝。

P 好。妳想走觀景路線還是最省時的路線？

C 快一點比較好，我的列車 30 分鐘後就要開了。

P 了解。就算現在路況不理想，應該還是能在 15 分鐘內到。

C 太好了，謝謝。到車站大概多少錢？

P 大概 12 歐元左右，不過確切價格要看到那裡時跳表跳多少錢。

C 可以麻煩你在西側入口放我下車嗎？我的列車在第五月臺。

P 沒問題。

C 可以刷卡嗎？

P 現金、刷卡都可以，我這有刷卡機。

C 謝謝。

P 不客氣。請繫好安全帶，放心享受這段路程。

旅遊字詞補給包

❶ **hail** [hel] *vt.* 招呼，呼叫；讚揚
After waiting for ten minutes, Amy finally managed to hail a ride.
等了十分鐘後，艾咪終於叫到車了。

❷ **hop in**　　（快速地）跳進（車子）
Hey, hop in! We're going to be late.
嘿，快上車！我們要遲到了。

❸ **via** [ˋvaɪə] *prep.* 經由
The news spread quickly via social media.
這個消息透過社群媒體迅速散播開來。

❹ **scenic** [ˋsinɪk] *a.* 風景優美的
The hotel room had a scenic view of the mountains.
這個飯店房間可看見美麗的山色。

❺ **route** [rut / raʊt] *n.* 路線
We took a different route home to avoid traffic.
我們為了避開交通堵塞而走了不同的路線回家。

❻ **platform** [ˋplæt͵fɔrm] *n.* 月臺
The platform was crowded with people during rush hour.
尖峰時段月臺上擠滿了人。

經典旅遊實用句

Can you drop me at [特定地點], please?
可以麻煩你在……放我下車嗎？

用途：跟計程車司機指定下車地點。
可以麻煩你在西側入口放我下車嗎？
Can you drop me at the west entrance, please?

at the main entrance	在主要入口處
near the food court	在美食街附近
at Gate B	在 B 大門

旅人私房筆記　　Hey! Hey! Taxi!

一起來看看幾個具有代表性的計程車：

Yellow Cab　美國紐約
黃色計程車是紐約的代表

Hackney carriage (black cab)　英國倫敦
以黑色哈尼克車型最為經典

takushii（タクシー）　日本
日式計程車車門會自動開關，
司機會穿西裝打領帶

Taxi Meter　泰國
計程車外觀常寫上「TAXI METER」，
但司機有時不跳表（尤其是對觀光客）

Try It Out! 試試看　　請依句意在空格內填入適當的字詞

從這裡到機場的最佳路線是什麼？
What's the best _____ to the airport from here?

Ans route

旅遊英語輕鬆聊　　參考答案請掃描使用說明頁上 QR code 下載

In your country, is it common to chat with taxi drivers, or do people usually stay quiet?

CH 04　Transportation・交通

147

Unit 36 Taking the Bus 搭公車

實境對話 GO！

M Mario（馬力歐）　　**J** Jessica（潔西卡）

Mario is on vacation in London and is asking Jessica, a local, for transportation advice.

M Excuse me. Could you help me, please?

J Sure. How can I help you?

M Which bus do I take if I want to go to Big Ben?

J You need to go to Westminster. Bus number 24 is your best bet.

M Thank you. Where is the bus stop?

J Just across the road. Bus stop B. Can you see it?

M Yes, I can. Where do I buy tickets? Can I pay the driver?

J You don't need a ticket. Just tap your contactless payment card, smartphone, or Oyster travel card when you board the bus. You don't need to tap again when you get off.

M Thank you for the information!

J You're welcome. If you're not sure where to get off, the driver can tell you when you're near Westminster.

M▶ Oh, I'm sure we'll know where to get off. We'll just keep an eye out for a big clock!

J▶ That's right. It's not far. It should only take ten to fifteen minutes to get there. The buses run every five minutes, so you won't need to wait long.

M▶ Excellent. Thanks so much.

J▶ My pleasure. I hope you enjoy seeing Big Ben!

馬力歐在倫敦度假，正向本地人潔西卡詢問如何搭車。

M▶ 不好意思，可以幫個忙嗎？

J▶ 可以啊，有什麼事？

M▶ 我要去看大笨鐘，該搭哪一路公車？

J▶ 你要去西敏寺方向。搭 24 路公車最方便。

M▶ 謝謝。公車站在哪裡？

J▶ 就在馬路對面，編號 B 的那個公車站。有看到嗎？

M▶ 有。票在哪裡買？我可以直接付錢給司機嗎？

J▶ 不用買票。上車時刷感應式付款卡、智慧手機，或是牡蠣卡都可以。下車時不用再刷。

M▶ 謝謝妳告訴我這些！

J▶ 不客氣。如果你不確定在哪站下車，司機會提醒你西敏寺快到了。

M▶ 哦，我們應該是知道在哪裡下車的。我們認準那個大鐘就對了！

J▶ 沒錯。此地離西敏寺沒多遠，大概十到十五分鐘就到了。公車每五分鐘一班，所以你們也不用等太久。

M▶ 太好了，非常感謝。

J▶ 不客氣，希望你們覺得大笨鐘好玩！

旅遊字詞補給包

❶ **transportation** [ˌtrænspəˈteʃən] *n.* 交通運輸（工具）（不可數）
The government wants to improve the transportation system in the country.
政府要改善國內的運輸系統。

❷ **best bet**　　最可靠安全的辦法
If you want to get downtown quickly, taking the MRT is your best bet.
如果你想快速到達市中心，坐捷運是最好的選擇。

❸ **tap** [tæp] *vt. & vi.* 輕觸；輕敲
Please tap the screen to start the application.
請輕觸螢幕以啟動應用程式。

❹ **contactless** [ˈkɑntæktləs] *a.* （支付等）感應式的；無接觸的
More and more stores now offer contactless payment options.
現在越來越多商店提供感應支付的選項。

❺ **payment** [ˈpemənt] *n.* 支付，付款
The store accepts various forms of payment, including credit cards and cash.
這間商店接受多種付款方式，包括信用卡和現金。

經典旅遊實用句

Which [交通工具] do I take if I want to go to [地點]?
如果我想去 [地點]，應該搭哪一班 [交通工具]？

用途：詢問要前往某個目的地時該搭什麼交通工具。
我要去看大笨鐘，該搭哪一班公車？
Which bus do I take if I want to go to Big Ben?

train	火車	get to the museum	到博物館
subway line	地鐵線	reach the tower	抵達塔樓
tram	電車	visit Oxford Street	造訪牛津街

旅人私房筆記

Oyster Card 是什麼？

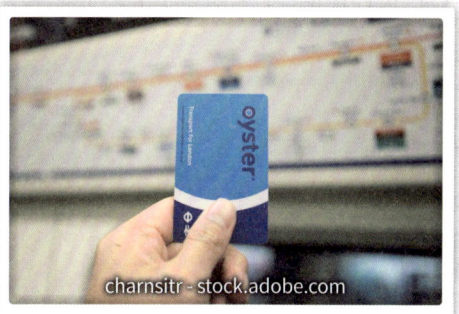

Oyster 卡可以用來搭乘以下交通工具：

Oyster Card　牡蠣卡
charnsitr - stock.adobe.com

是倫敦的大眾交通儲值卡，方便民眾和旅客搭乘倫敦的各種交通工具

Tube
地下鐵

Bus
公車

Overground
倫敦地上鐵

National Rail
國鐵（部分路線）

River Bus
泰晤士河快線

DLR
輕軌

如果你要在倫敦旅遊數天，Oyster card 是非常推薦的交通付款方式！

CH 04

Transportation・交通

Try It Out! 試試看　請依句意在空格內填入適當的字詞

① 如果你想找間不會踩雷的餐廳，那間義大利餐廳是最好的選擇。
If you're looking for a restaurant that won't let you down, that Italian place is your _____ _____.

② 我們已收到您的付款，並將盡快處理您的訂單。
We have received your _____ and will process your order ASAP.

Ans　① best bet　② payment

旅遊英語輕鬆聊　參考答案請掃描使用說明頁上 QR code 下載

Have you ever asked a stranger for directions while traveling?

151

Unit 37 Getting on the Wrong Bus
公車迷途記

實境對話 GO！

D Dave（戴夫）　　M Mina（米娜）

Mina is going to meet her friend Dave. He is curious about where she is at the moment.

D Hi, Mina, are you almost here?

M Yeah, I suppose. I've been on the bus for roughly 30 minutes.

D Cool. I'm looking forward to seeing you! Do you know where you are now?

M It beats me. There are lots of trees along the way; I think we just left the city.

D I don't live that far from downtown. Are you positive you're on the right bus?

M I think so. I got on the 590 bus. Hold on, there is a stop coming up.

(The driver calls out "Grand Memorial Park" as he makes a stop.)

M I'm at Grand Memorial Park.

D Oh, no. That's on the north side of the city. I actually live on the south side.

M What? You're kidding, right?

D No, I'm afraid you're heading in the wrong direction.

M What should I do?

D That bus makes a loop through the city, so you can ride it all the way around. But you'll get here faster if you transfer to another bus.

M OK, I'll try to catch a southbound bus at the next stop. Sorry I'm running so late.

D Don't worry about it. Talk to you soon.

> 米娜要去見她的朋友戴夫。他很好奇米娜現在到哪裡了。

D 嗨，米娜，妳快到我這了嗎？

M 應該是吧，我想。我上公車大概有三十分鐘了。

D 太棒了。我好期待見到妳喔！妳知道自己現在身在何處嗎？

M 你把我問倒了。沿途有好多樹；我覺得我們這輛公車才剛離開市區。

D 我家離市區沒那麼遠。妳確定妳有搭對公車嗎？

M 應該是吧。我搭上 590 公車。等等，接近站牌了。

(公車駕駛靠站時喊出「紀念公園站」。)

M 我在紀念公園。

D 不是吧。那可是在市區的北邊。我其實是住在南側。

M 什麼？你在跟我開玩笑對吧？

D 並沒有，恐怕妳搭錯方向了。

M 那我該怎麼辦呢？

D 那輛公車會環繞這座城市開，所以妳一路搭到底就是了。但要是妳轉乘別輛公車就會更快到我這兒。

M 好，我會試著下一站改搭往南的公車。抱歉我拖到這麼晚。

D 別擔心。見面聊。

旅遊字詞補給包

❶ be curious about...　　對……感到好奇
Kids are curious about everything they see.
小朋友對看到的一切都覺得好奇。

❷ **roughly** [ˈrʌflɪ] *adv.* 大約，粗略地（可修飾數字）
It will take roughly three hours to get to New York from here.
從這裡到紐約大約需要三個小時。

❸ **downtown** [ˌdaʊnˈtaʊn] *n.* 市中心，鬧區 & *adv.* 在／向市中心
It's hard to find a parking space downtown on the weekend.
週末很難在市區找到停車位。

❹ **positive** [ˈpɑzətɪv] *a.* 肯定的，確信的
Jeremy is positive that he paid me back, but I don't remember him doing so.　傑若米很肯定他已把錢還給我，但我卻不記得他有還。

❺ **come up**　　接近，靠近
A big test is coming up next week, so the students are all studying hard.　下週將有一次大考，所以學生們都在為此努力研讀。

❻ **head** [hɛd] *vi.* 向……去，朝……前進
We must catch the southbound train heading to Chiayi at 1:04 p.m.
我們必須趕上下午一點零四分那班開往嘉義的南行列車。

❼ **transfer** [trænsˈfɝ] *vi.* 轉乘，改乘　→　三態為：transfer, transferred [trænsˈfɝd], transferred
My parents' house is in a remote area, so I need to transfer to two different buses.
我老家很偏僻，所以我需要轉搭兩班不同的公車才能到。

❽ **run late**　　遲到；晚了
I'm running late for my interview. Can you give me a ride?
我面試快要遲到了。你可以順道載我一程嗎？

經典旅遊實用句

Are you positive you're on the [交通工具]?
你確定你搭的是……嗎？

用途：用於確認對方是否選擇正確交通工具，常在迷路時使用。

你確定你有搭對公車？
Are you positive you're on the right bus?
　　　　　　　　　　　　　　correct train　　　　對的火車
　　　　　　　　　　　　　　right subway line　　正確的地鐵路線

旅人私房筆記　　　　**什麼是 Loop Bus（環狀公車）？**

常見的 loop bus 有：

- **市區接駁車（Downtown Loop）**：
繞行市中心、商圈、重要景點等

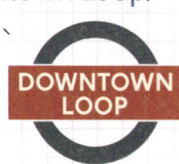

Loop Bus　　環狀公車

Rafael Ben-Ari - stock.adobe.com

loop 在英文中是「環狀、循環」的意思。當一輛公車「makes a loop」，代表它從起點出發，沿著一定路線繞一圈，最後又回到起點。

- **校園巴士（Campus Loop）**：
學校內多點循環

- **觀光巴士（Sightseeing Loop）**：
為觀光客設計，停靠熱門景點

CH 04　Transportation・交通

Try It Out! 試試看　　請從下欄選出最適合的句子來完成該句

____ ❶ All buses going downtown are running an hour late

____ ❷ Whether you are a man or a woman,

____ ❸ I am pretty positive that Becky

A. we are all curious about the thoughts of the opposite sex.

B. will be on my side of this argument.

C. because of heavy traffic.

Ans　❶ C　　❷ A　　❸ B

旅遊英語輕鬆聊　　參考答案請掃描使用說明頁上 QR code 下載

Have you ever missed your stop because you weren't paying attention?

155

Notes

Chapter 05

Sightseeing
觀光

Unit 38
Touring in the UK
英國旅遊趣（詢問旅客資訊站）

Unit 39
Buying Tickets
買票

Unit 40
Entering an Amusement Park
驗票入遊樂園

Unit 41
Fun at the Amusement Park
遊樂園小約會

Unit 42
Touring a Museum
參觀博物館

Unit 43
Taking in a Theater Show
一起去看戲

Unit 44
Preparing for a Skiing Trip
準備滑雪趣

Unit 45
Taking Pictures
拍照

Unit 38 Touring in the UK
英國旅遊趣（詢問旅客資訊站）

實境對話 GO！

L Lucy（露西）　　**C** Consultant（顧問）

A tourist walks up to a tourism kiosk in a London park.

L Excuse me. This is my first time in London. Can you tell me something about this city, please?

C Sure. London is the capital of the UK. It is one of the world's largest financial centers, and there are many places worth visiting.

L Do you have any one-day sightseeing tours around London available?

C We don't have any one-day tours around London. All of our one-day tours are day trips outside of the city. We do have a three-hour tour that visits Big Ben, Buckingham Palace, Westminster Abbey, and so on.

L The three-hour tour sounds interesting! How much does it cost?

C Only 15 pounds. And with the weather as good as today, you'll be able to take some beautiful shots of the abbey and palace.

L Sounds great. When does it start?

C We have tours every hour on the hour, which means the next one will start at 11.

L OK. So I still have a half hour to grab something to eat. Can I also book a one-day tour for tomorrow? I want to see more of England.

C Certainly! We have one that visits Stonehenge, Bath, and Windsor Castle. It costs 90 pounds and includes a meal at a local pub.

L Cool! I'll go on that one.

C Here's your reservation number. Hope you enjoy your time in England.

一名遊客走到倫敦公園的旅遊資訊服務亭。

L 不好意思。這是我第一次到倫敦。可以請你稍微介紹一下這個城市嗎？

C 當然。倫敦是英國的首都。它是世界最大的金融中心之一，而且也有很多值得參觀的景點。

L 你們有任何倫敦一日遊的觀光旅程嗎？

C 我們沒有倫敦一日遊的行程。我們所有的一日遊都是在倫敦外的一日行程。但我們有參觀大笨鐘、白金漢宮及西敏寺等景點的三小時導覽。

L 三小時導覽聽起來很有趣！大概要多少錢呢？

C 只要十五英鎊。而且像今天這樣的好天氣，您能夠拍到西敏寺及白金漢宮的一些美照。

L 聽起來不錯。導覽什麼時候開始呢？

C 我們在每個小時的整點都有導覽，也就是說下一場會在十一點鐘開始。

L 好的。所以我還有半小時的時間去找點食物來吃。我可以也預訂明天的一日遊嗎？我想多參觀英國。

C 沒問題！我們有一個去參觀巨石陣、巴斯以及溫莎城堡的行程。價格是九十英鎊且內含在當地酒吧用餐。

L 太好了！我要去這個。

C 這是您的預約號碼。希望您在英國玩得愉快。

旅遊字詞補給包

❶ walk up to... 向……走過去
Andy walked up to his favorite singer and asked for a picture.
安迪走向他最喜歡的歌手並要了一張合照。

❷ **kiosk** [ˈkiˌɑsk] *n.* (賣報紙、票券、查資訊等) 服務亭 / 機
You can buy tickets for the museum at the information kiosk near the entrance.
你可以在博物館入口附近的資訊票亭購買門票。

❸ **capital** [ˈkæpətl̩] *n.* 首都
Paris is the capital of France.
巴黎是法國的首都。

❹ **financial** [faɪˈnænʃəl] *a.* 金融的
It's important to have a good financial plan for the future.
為以後的人生訂定一個妥善的財務計畫是很重要的。

❺ **be worth + N/V-ing**　　值得……
This song is worth listening to again and again.
這首歌值得一聽再聽。

❻ **shot** [ʃɑt] *n.* 照片
I took some good shots of the lake at dawn.
我於黎明時分拍了一些湖泊的美麗照片。

❼ **local** [ˈlokl̩] *a.* 當地的，本地的
Daniel took his car to a local garage to get it fixed.
丹尼爾將車子送到當地的一家修車廠維修。

經典旅遊實用句

Do you have any [天數] [行程] around London available?
你們有任何倫敦 [天數] 的 [行程] 嗎？

用途：詢問是否有特定地點的導覽行程可供預訂。

你們有任何倫敦一日遊的觀光旅程嗎？
Do you have any one-day　sightseeing tours around London available?
　　　　　　　　half-day　guided tours　　導覽行程
　　　　　　　　半日
　　　　　　　　multi-day　bus tours　　　巴士觀光行程
　　　　　　　　多日　　　 walking tours　徒步旅遊行程

旅人私房筆記

英國旅遊小字典

Bath
巴斯

Big Ben
大笨鐘

Westminster Abbey
西敏寺

Buckingham Palace
白金漢宮

Stonehenge
巨石陣

CH 05 Sightseeing・觀光

Try It Out! 試試看　請依句意在空格內填入適當的字詞

愛黛兒的演唱會值得一去。
Adele's concert is _____ going to.

Ans worth

旅遊英語輕鬆聊　參考答案請掃描使用說明頁上 QR code 下載

Would you rather visit a castle, a church, or a pub in England?

Unit 39 Buying Tickets
買票

實境對話 GO！

S Susannah（蘇珊娜）　　**A** Albert（亞伯特）

Susannah is buying tickets from Albert, who works at a museum.

S Good afternoon. I would like two adult tickets and one child ticket for the museum, please.

A How old is the child?

S She's ten.

A That's fine—child tickets are for all children under thirteen. Standard adult **admission** is $20. Child admission is $10.

S Does that include **entry** to the special Andy Warhol **exhibition**?

A Yes, it does. You'll find that in **Gallery** 2 on the second floor.

S Are there any guided tours available?

A Unfortunately, you just missed the final guided tour of the day. However, we do offer **audio** guides, which come in **multiple** languages. They're priced at $5 each.

S OK, I'll take three of those. We'd like to know as much about the exhibitions as possible!

A So, that will be $65 in total, please.

S Can I pay by card?

A Yes, you can.

S Great! Here's my card. What time does the museum close?

A It closes at 6:30 p.m., so you've still got a couple of hours to see everything. Here's your card, and here are your headphones along with **instructions** on how to use them.

S Perfect. Thanks so much!

A You're welcome. Enjoy your visit!

蘇珊娜正在向博物館職員亞伯特購票。

S 午安，我要兩張全票和一張兒童票。

A 小朋友幾歲？

S 十歲。

A 那可以 —— 13 歲以下都可以買兒童票。全票 20 美元，兒童票 10 美元。

S 這票可以看安迪‧沃荷特展嗎？

A 可以。展覽在二樓的第二展廳。

S 有導覽嗎？

A 很不巧，今天最後一場導覽剛結束。不過我們有提供多種語言的語音導覽，每個租金 5 美元。

S 好，我租三個。我們想盡量了解展覽的細節。

A 這樣總共是 65 美元。

S 可以刷卡嗎？

A 可以。

S 太好了，卡在這裡。博物館幾點打烊？

A 晚上六點半打烊，所以你們還有幾個小時可以參觀。來，卡還您，這是語音導覽耳機以及使用說明。

S 很好，非常感謝！

A 不客氣，祝您參觀愉快！

旅遊字詞補給包

❶ **admission** [ədˈmɪʃən] *n.* 入場費；入會
The admission fee for the exhibition is NT$200.
這場展覽的入場券是新臺幣兩百元。

❷ **entry** [ˈɛntrɪ] *n.* 進入
Please show your ticket for entry to the concert hall.
請出示您的入場券才能進去音樂廳。

❸ **exhibition** [ˌɛksəˈbɪʃən] *n.* 展覽會
The exhibition will run until the end of next month.
這個展覽將持續到下個月底。

❹ **gallery** [ˈgælərɪ] *n.* 陳列館；畫廊
Anita dreams of one day having her own art gallery.
艾妮塔夢想有一天能擁有自己的藝廊。

❺ **audio** [ˈɔdɪˌo] *a.* 聲音的
I prefer listening to audio books when I'm commuting.
我通勤時比較喜歡聽有聲書。

❻ **multiple** [ˈmʌltəpl̩] *a.* 多重的，眾多的
The website offers multiple language options.
這個網站提供多種語言選項。

❼ **instruction** [ɪnˈstrʌkʃən] *n.* (機器等) 使用說明 (常用複數)
If you have any questions, please refer to the user instructions.
如果您有任何疑問，請參考使用說明。

經典旅遊實用句

I would like [數量] [票種] tickets, please.
我想要 [數量] [票種]。

用途：購票時最需要用到的數量及票種需求一次講清楚。

我要兩張全票和一張兒童票。
I would like <u>two</u> <u>adult</u> tickets and <u>one</u> <u>child</u> ticket for the museum.

<u>student</u>	學生票
<u>senior</u>	敬老票
<u>group</u>	團體票

旅人私房筆記　　　　各式常見導覽

guided tour
導覽人員帶領解說（通常有固定時段）

audio guide
語音導覽機，可選語言

virtual tour
虛擬導覽，有些博物館提供虛擬模式或是線上參觀

interactive exhibits
互動展覽，適合親子遊客

CH 05 Sightseeing・觀光

Try It Out! 試試看　　請依句意在空格內填入適當的字詞

那間展覽室一次限定五十人入內參觀。
_____ to that exhibit room is limited to 50 people at a time.

Ans　Entry

旅遊英語輕鬆聊　　參考答案請掃描使用說明頁上 QR code 下載

If you were visiting this museum, which would you prefer: a guided tour or an audio guide? Why?

165

Unit 40 Entering an Amusement Park
驗票入遊樂園

實境對話 GO！

P Park Staff Member（園區工作人員） **L** Leo（里歐）

Leo walks up with his family to the entrance of an amusement park.

P Welcome to King's Point Amusement Park. May I see your tickets, please?

L I have **digital** tickets on my phone that I bought online.

P Please raise your QR code to the **scanner**. I see you have tickets for two adults and one child. Is this your child right here?

L Yes, she is.

P Sorry for asking, but is she 12 or under? The age **limit** for children's tickets is 12.

L Oh, no. She just turned 13 a few days ago. She was 12 when I booked the tickets, though.

P This is a **tricky** situation, indeed. I'll make a quick call to my manager to see if we can make any **adjustments** for this special case. Otherwise, I'm afraid you'll have to pay for a full-price ticket.

L I understand.

(The staff member makes a call to her manager to explain the situation and returns to the entrance.)

P I have bad news. It appears that since your daughter is 13 on the day of entry to the park, she'll need an adult ticket.

L I see. Thanks for calling to check. How much will it be for three adult tickets?

P Your total will **come to** $150. I can help you **refund** her original ticket.

L OK. Here you go.

P Thank you and enjoy your day at King's Point Amusement Park.

> 里歐和他的家人走向遊樂園的入口。

P 歡迎來到歡樂王國遊樂園。麻煩能讓我看一下你們的票嗎？

L 我的手機裡有在網路上購買的電子票券。

P 請將您的 QR 碼放到掃描器上。我看您持有的票是兩張成人票及一張兒童票。這是您的孩子嗎？

L 是的。

P 冒昧詢問，她十二歲以下嗎？兒童票的年齡上限是十二歲。

L 噢，糟糕。她前幾天才剛滿十三歲。不過我訂票時她是十二歲。

P 這的確是個棘手的狀況。我得趕快打電話給我的經理，看我們是否能為這個特殊狀況做任何調整。否則您恐怕得付全票價。

L 我明白了。

（該工作人員打電話給她的經理說明情況並返回入口。）

P 我有個壞消息。看來您女兒在入園的這天是十三歲，所以她需要一張成人票。

L 了解。謝謝您打電話確認。請問三張成人票要多少錢呢？

P 總共是一百五十美元。我可以幫您退她原來的票。

L 好的。這是一百五十美元。

P 謝謝您，祝你們今天在歡樂王國遊樂園玩得愉快。

旅遊字詞補給包

1. **digital** [ˈdɪdʒətḷ] *a.* 數位的
John bought a digital watch for his mother.
約翰買了一支電子錶給他媽媽。

❷ **scanner** [ˈskænɚ] *n.* 掃描器
At the library, you can use the public scanner to scan book materials.
在圖書館裡，你可以用公用掃描器來掃描書籍資料。

❸ **limit** [ˈlɪmɪt] *n.* 限制；極限
I set a limit on how much I can spend on food every month.
我每個月會限制自己的食物開銷。

❹ **tricky** [ˈtrɪkɪ] *a.* 棘手的，難處理的
Emily came up with a clever solution to the tricky problem.
愛蜜莉想出一個聰明辦法來解決那個棘手問題。

❺ **adjustment** [əˈdʒʌstmənt] *n.* 調整
After the software update, some minor adjustments to the settings were needed.
軟體更新後，需要對設定進行些微的調整。

❻ **come to + 金額**　　總共為 / 共計……（金額）
Lillian's credit card bill last month came to NT$8,000.
莉莉安上個月的信用卡費總共是新臺幣八千元。

❼ **refund** [rɪˈfʌnd] *vt.* 退還（款項）
The handle on the purse Judy had just bought broke off, so she asked the store to refund her money.
茱蒂剛買的那個皮包的提把斷了，所以她去店裡要求退錢。

經典旅遊實用句

May I see your [票 / 證明], please?
我可以看你的 [票 / 證明] 嗎？

用途：服務人員常用的基本禮貌詢問句，用於索取票券、證件、護照等。
能讓我看一下你的門票嗎？
May I see your tickets, please?

ID	身分證
reservation	預約
passport	護照
confirmation email	確認信

旅人私房筆記

遊樂設施的英文怎麼說？

roller coaster
雲霄飛車

tea cups
咖啡杯

Ferris wheel
摩天輪

merry-go-round
旋轉木馬

pirate ship
海盜船

CH 05 Sightseeing・觀光

Try It Out! 試試看 請依句意在空格內填入適當的字詞

喬許為了拍自己的小孩買了一臺數位相機。
Josh bought a ＿＿＿＿＿ camera in order to take pictures of his kids.

Ans　digital

旅遊英語輕鬆聊 參考答案請掃描使用說明頁上 QR code 下載

If you were visiting an amusement park, would you prefer to buy tickets online or at the entrance? Why?

Unit 41 Fun at the Amusement Park
遊樂園小約會

實境對話 GO！

I Ivy（艾薇）　　**K** Kevin（凱文）

Kevin and his new American girlfriend are at a large theme park for the day.

I We've already gone on the swinging-boat ride six times! Let's try something else.

K But there's almost no line.

I It's so hot out today. We should go on the log ride to cool down a bit.

K All right, but if the line looks long, I might give up.

I You are hungry, aren't you?

K Sorry, I can get a bit moody when I'm hungry.

I As do most people. After this ride, we'll get a snack. I love the food at places like this.

(After they go on the log ride, they head to the area with all of the food vendors.)

K That was a great ride! I'm glad I put my phone in a bag, but I wish I'd brought a change of clothes. What should we eat?

I I want a deep-fried Oreo that I saw earlier.

K Starting with **dessert**?

I Welcome to America.

K Sounds good to me, but I also want some real food.

I How about corn dogs and some **curly** fries?

K Perfect. Let's get those and a chocolate churro afterwards.

I A second dessert? Well, you know I can't say no to chocolate.

K It will be my treat.

凱文和他的新美國女友在一個大遊樂園裡一日遊。

I 我們已經坐了六次海盜船了！咱們來試試別的吧。

K 但是這裡幾乎沒有人排隊耶。

I 今天外面好熱。我們應該去坐急流獨木舟涼快一下。

K 好吧，不過如果看起來很多人排隊的話，我可能就不玩了。

I 你餓了，對吧？

K 抱歉，我肚子餓的時候情緒會有一點差。

I 大部分的人都會這樣啊。搭完這趟後，我們去吃點東西。我超愛這種地方的食物。

（搭完急流獨木舟之後，他們走到食物攤販聚集的地方。）

K 剛剛真是太好玩了！真慶幸我有把手機放在袋子裡，不過要是我有帶替換的衣服就好了。我們該吃什麼好呢？

I 我想吃剛才看到的炸 Oreo 餅乾。

K 從甜點開始嗎？

I 歡迎來到美國。

K 聽起來不錯啦，不過我也想要來點真正的食物。

I 炸熱狗和捲捲薯條怎麼樣？

K 好極了。咱們就點這幾種，之後再買根巧克力吉拿棒好了。

I 第二道甜點？嗯，你知道我無法拒絕巧克力的。

K 我請客哦。

旅遊字詞補給包

三態為：swing, swung, swung

❶ **swing** [swɪŋ] *vi.* & *vt.* (使)(前後)擺動 / 搖擺；揮動(器物 / 拳頭)打擊
The clothes I hung out to dry earlier are swinging in the wind.
我稍早晒在外頭的衣物隨風擺盪。

❷ **vendor** [ˋvɛndɚ] *n.* (尤指街頭的)小販
Leah bought some hot dogs from the vendor before the baseball game began.
莉亞在棒球賽開始前向小販買了一些熱狗。

❸ **deep-fried** [ˋdip͵fraɪd] *a.* 油炸的；油煎的
Eating too much deep-fried food is not good for your health.
吃太多油炸食物有害健康。

❹ **dessert** [dɪˋzɝt] *n.* (飯後的)甜點，甜品
There were many kinds of dessert to choose from on the menu.
菜單上有許多甜點可供選擇。

❺ **curly** [ˋkɝlɪ] *a.* 捲曲的
Anne's curly hair always ends up in a mess after the wind blows.
小安的捲髮經風吹後總會亂成一團。

經典旅遊實用句

How about [食物名稱]?　　[食物名稱] 怎麼樣？

用途：提出食物建議，適合討論要吃什麼的時候使用。

炸熱狗和捲捲薯條怎麼樣？
How about corn dogs and some curly fries?

　　　　　hamburgers　漢堡
　　　　　ice cream　　冰淇淋
　　　　　French fries　薯條
　　　　　pizza　　　　披薩

旅人私房筆記　　遊樂園餓了吃什麼？

我們一起來看看在遊樂園可以買到什麼？

popcorn
爆米花

fried chickens
炸雞

cotton candy
棉花糖

a hot dog
熱狗

churros
吉拿棒

Try It Out! 試試看　　請依句意在空格內填入適當的字詞

❶ 這是我吃過最好吃的炸豆腐。
　　This _____ tofu is the best I've ever had.

❷ 吉姆揮棒擊球卻落空了。
　　Jim _____ the bat at the ball but missed it.

❸ 市場的小販免費送我這些蔥。
　　The _____ at the market let me have these green onions for free.

Ans　❶ deep-fried　❷ swung　❸ vendor

旅遊英語輕鬆聊　　參考答案請掃描使用說明頁上 QR code 下載

Your friend wants to go on a scary ride, but you're a bit scared. What do you say to find agreement?

CH 05

Sightseeing・觀光

173

Unit 42 Touring a Museum
參觀博物館

實境對話 GO !

J Jennifer（珍妮佛）   **A** Alex（艾力克斯）   **C** Chris（克里斯）

Jennifer and her friends are traveling together and have decided to visit a museum for the day.

J We finally made it to the **Metropolitan** Museum of Art. Do you guys want to take a tour?

A I don't know that much about art, so it would be nice to have a guide to tell me what I'm looking at.

C That's true, but I don't want to **get stuck with** a huge group of people. We'll be **in a rush** and won't get to stop and **appreciate** the artwork.

A Are there other choices?

J **According to** the sign over there, we've got three choices. There's a tour leaving in 10 minutes, we can get audio guides, or we could just use the app.

C Anything but the tour.

A Well, I'd prefer the tour, but let's get the app and try it.

J I forgot my phone at the hostel, so I'll just get the audio guide.

C I don't want to spend money on that, so I'll try the app.

A Here's an idea. Why don't we just separate? I'll take the tour, and we can meet back here in two hours.

C Perfect! This way, we're all happy.

J OK. Have fun and see you guys soon.

> 珍妮佛和朋友一起旅行，他們決定今天要去參觀一間博物館。

J 我們終於到大都會藝術博物館了。你們大家想要參加導覽團嗎？

A 我對藝術不是那麼了解，所以如果可以有一個導覽員告訴我，我正在看的是什麼的話會很不錯。

C 是這樣沒錯，但是我不想和一大群人一起行動。那樣的話我們會很匆忙，沒有辦法停下來好好欣賞藝術作品。

A 有別的選擇嗎？

J 依那邊的告示牌來看，我們有三種選擇。有一批導覽團將在十分鐘後出發，我們也可以使用語音導覽，或是用應用程式就好。

C 我除了導覽團都可以。

A 嗯，我比較想參加導覽團，不過咱們來下載應用程式試試看吧。

J 我把手機忘在青年旅舍了，所以我會用語音導覽。

C 我不想花錢，所以我會試試應用程式。

A 我有一個主意。我們何不分開走呢？我加入導覽團，然後我們兩個小時後回到這裡碰面。

C 太棒了！這麼一來我們大家都滿意了。

J 好的。你們玩得開心點，大家待會見囉。

旅遊字詞補給包

❶ **metropolitan** [ˌmɛtrəˈpɑlətn̩] *a.* 大都市的
Kevin works in the metropolitan area.
凱文在都會區工作。

❷ **get / be stuck with sb** 擺脫不了某人，甩不掉某人
Since the rain doesn't seem like it will stop soon, we're stuck with each other for now.
這場雨看起來短時間內不會停，這下我們只得暫時待在一起啦。

❸ **in a rush**　　趕時間；匆忙 / 很快地
Bryce tends to do everything in a rush even though he could just take it easy.　布萊斯做事習慣匆匆忙忙的，即使他大可放輕鬆點。

❹ **appreciate** [əˋpriʃɪˌet] *vt.* 欣賞；賞識；鑑賞
Only a few people could appreciate May's talents.
只有少數幾個人懂得賞識小梅的才華。

❺ **according to...**　　據……所示，按……所說
According to *The New York Times*, one of the best-sellers of 2005 was Dan Brown's *The Da Vinci Code*.
根據《紐約時報》所寫，丹・布朗的《達文西密碼》是 2005 年的暢銷書之一。

經典旅遊實用句

Do you guys want to [做某事]?　你們大家想要……嗎？

用途：語氣輕鬆地提議活動或詢問意願。

你們大家想要加入導覽團嗎？
Do you guys want to take a tour?
　　　　　　　　　go shopping　　買東西
　　　　　　　　　grab lunch　　　吃午餐
　　　　　　　　　take a break　　休息一下

旅人私房筆記　　　　想看什麼展？

帶你認識各種展覽的英文怎麼說：

photography exhibition
攝影展

art exhibition
藝術展

ancient artifacts exhibition
古文物展

fashion exhibition
服裝展

science exhibition
科學展

Try It Out! 試試看　請依句意在空格內填入適當的字詞

潔西卡卡在那個難解的數學題，沒法把功課做完。
Jessica got _____ _____ that tricky math problem and couldn't finish her homework.

Ans stuck with

旅遊英語輕鬆聊　參考答案請掃描使用說明頁上 QR code 下載

> If you only had time to see one exhibition in a museum, what would you choose and why?

Unit 43 Taking in a Theater Show
一起去看戲

實境對話 GO！

🅑 Brett（布雷特）　🅢 Stacey（史黛西）

Brett and Stacey are on vacation in Las Vegas and want to see a theater performance.

🅑 Wow. The Las Vegas Strip is such a breathtaking area!

🅢 I know. There is so much to do. I'm not sure where to begin.

🅑 You said you want to see a performance while we are here, didn't you?

🅢 Yeah, let's see what kind of shows are happening this weekend.

(The couple walk over to a ticket vendor to look at the advertisements.)

🅑 Whoa, look at all these options. Do you want to see a magic show, a musical concert, a circus performance, stand-up comedy, or something more traditional?

🅢 Oh, look! There is a SpongeBob SquarePants musical!

🅑 I used to love that cartoon. I bet that would be a lot of fun.

🅢 Let's see if there are any seats still available.

(Stacey asks the ticket teller about show times and ticket prices.)

🅢 It looks like the only tickets left are for a show tonight at 7:30 p.m.

🅑 Really? But that's just two hours from now.

S> That's plenty of time. We can get some food and then change clothes at the hotel.

B> All right. Let's do it!

S> This is going to be a **memorable** night!

布雷特和史黛西來到拉斯維加斯度假,他們想看一齣劇場演出。

B> 哇。賭城大道真是個令人嘆為觀止的地方!

S> 就是說啊。有太多事可以做了。我都不知道該從哪裡開始。

B> 妳說妳想趁我們在這裡的時候看場表演,對吧?

S> 對啊,來看看這個週末有哪些好戲上演。

(情侶兩人走到售票亭旁瀏覽廣告。)

B> 哇,看看這些選擇。妳想看場魔術秀,還是音樂會、馬戲團表演、脫口秀,或者來點比較傳統的?

S> 喔,你看!有一場海綿寶寶音樂劇!

B> 我以前好愛那卡通。我敢肯定那一定會很有趣。

S> 來看看還有沒有空位。

(史黛西向售票員詢問時刻表和票價。)

S> 看來只剩下今晚七點半的秀還有票。

B> 真的嗎?但這樣只剩兩個鐘頭了。

S> 時間綽綽有餘。我們可以買點食物然後回飯店換衣服。

B> 好。就這麼辦吧!

S> 這將會是個難忘的夜晚!

旅遊字詞補給包

❶ **performance** [pɚˋfɔrməns] *n.* 表演
The band gave a wonderful performance tonight.
這個樂團今晚的演出非常精彩。

❷ **breathtaking** [ˋbrɛθˌtekɪŋ] *a.* 令人讚嘆的
The breathtaking view is the main reason we live in this apartment building. 我們住這棟公寓大樓的主因,是因為這裡有令人嘆為觀止的景色。

❸ **advertisement** [ˌædvɚˈtaɪzmənt] *n.* 廣告 (= ad)
My company will place an advertisement in the newspaper for a secretary next week.
我的公司下星期將在報上刊登徵聘祕書的廣告。

❹ **concert** [ˈkɑnsɚt] *n.* 音樂會，演唱會
The concert was attended by thousands of fans.
那場音樂會有數以千計的樂迷參加。

❺ **circus** [ˈsɝkəs] *n.* 馬戲團
The school principal invited two circus performers to entertain the students.
校長請了兩位馬戲團表演者來娛樂學生。

❻ **available** [əˈveləbl̩] *a.* 可得到的，可買到的
The singer's latest album is now available.
該歌手的最新專輯已經上市了。

❼ **memorable** [ˈmɛmərəbl̩] *a.* 令人難忘的
Rita had a memorable experience in Europe.
莉塔在歐洲有一段令她難忘的經歷。

經典旅遊實用句

Let's see if there are any [票 / 時段] still available.
來看看還有沒有票 / 空位。

用途：表達想查詢是否還有票，或是否還有空位。

來看看還有沒有空位。
Let's see if there are any seats still available.

tickets	票
tours	行程名額
time slots	時段

旅人私房筆記　　　Vegas 的飯店也太狂！

The Venetian　威尼斯人酒店
有室內運河可以搭乘「貢多拉」小船。

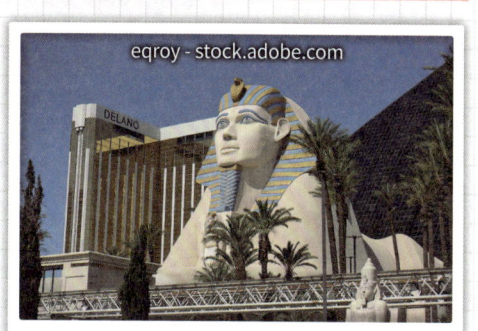

Luxor Hotel　盧克索酒店
是金字塔造型，連電梯也是斜的。

Paris Las Vegas　巴黎酒店
飯店前有縮小版的巴黎鐵塔，還可以登上去看夜景。

New York-New York Hotel & Casino　紐約 - 紐約酒店
模擬紐約市風景，有自由女神像、帝國大廈縮影，還有迷你雲霄飛車。

Try It Out! 試試看　　請依句意在空格內填入適當的字詞

你不准在這面牆上張貼任何廣告。
You're not allowed to post any _____ on this wall.

　　　　　　　　　　　　　　　　　Ans　advertisements

旅遊英語輕鬆聊　　參考答案請掃描使用說明頁上 QR code 下載

If you could watch any kind of live performance in Las Vegas, what would it be and why?

CH 05　Sightseeing・觀光

Unit 44 Preparing for a Skiing Trip
準備滑雪趣

實境對話 GO!

V Vincent（文森）　　**E** Emma（艾瑪）

Vincent and Emma are sitting in their hotel room at a ski resort in Japan.

V Hey, Emma, are you excited to ski tomorrow?

E Certainly! I can't wait to hit the slopes again. It's been a year.

V That's a long time! I'm actually a bit nervous to ski since it's my first time.

E It's going to be fun; trust me. The snow in Japan is super dry and lighter than anywhere else.

V Wonderful! I can't wait to experience it for myself.

E Have you checked all your skiing gear?

V Yes, I made sure to pack everything.

E Nice. Oh, what time is the skiing lesson you booked?

V It's at 10 a.m. I'll have enough time to get ready and grab breakfast with you before the lesson.

E Perfect! Also, do you know if you need to rent any skiing equipment?

V My lesson comes with rental gear, so the ski coach will help me pick out skis and ski poles at the lodge tomorrow morning.

E Great, that's sorted then. After your lesson, we can go on some slopes together.

V That would be fun. I'm so glad you invited me on this trip.

E Same. Let's make the most of it!

文森和艾瑪正坐在日本滑雪勝地的飯店房間裡。

V 嘿，艾瑪，妳期待明天去滑雪嗎？

E 當然！我等不及再次滑雪了。距離上次已經一年了。

V 那很久耶！我其實有點緊張因為這是我第一次滑雪。

E 會很有趣的，相信我。日本的雪超級乾燥，而且比其他地方都還要輕。

V 太棒了！我等不及親自體驗了。

E 你檢查過所有滑雪裝備了嗎？

V 有，我確定都打包好所有東西了。

E 很好。喔，你預約的滑雪課程是幾點？

V 早上十點。上課前，我會有充足的時間做準備並和妳趕緊吃個早餐。

E 完美！還有，你知道你需不需要租借任何滑雪器材嗎？

V 我的課程附有租借的裝備，所以明天早上滑雪教練會在度假小屋幫我挑選滑雪板和滑雪杖。

E 很好，那就解決了。你下課之後，我們可以一起去滑雪。

V 那會很有趣的。我很高興妳邀請我參加這次的旅行。

E 彼此彼此。我們玩個盡興吧！

旅遊字詞補給包

❶ **hit the slopes**　　去滑雪
slope [slop] *n.* 山坡
Tom can't wait for winter to come so he can hit the slopes.
湯姆等不及冬天的到來，這樣他就可以去滑雪了。

❷ **gear** [gɪr] *n.* 裝備 (不可數)
We need to pack our camping gear for the weekend trip.
我們需要打包週末旅行的露營裝備。

❸ **grab** [græb] *vt.* 匆忙地做某事 ⟶ 三態為：grab, grabbed, grabbed
Charlie grabbed some breakfast and went to school.
查理匆匆吃了點早餐就去上學了。

❹ **equipment** [ɪˋkwɪpmənt] *n.* 用具，器材 (不可數)
The gym has a wide range of exercise equipment.
這間健身房有多種運動器材。

❺ **come with...** 附有……
All main meals come with bread, soup, and a drink.
所有主餐均附有麵包、湯和飲料。

❻ **pick out... / pick... out** 挑選……
Gary picked out a beautiful dress for Elva as her birthday gift.
蓋瑞挑了一件漂亮的洋裝給艾娃當生日禮物。

❼ **lodge** [lɑdʒ] *n.* 山中小屋，度假小屋
The national park has several lodges where visitors can stay overnight.
這個國家公園有數間供遊客過夜的小屋。

❽ **sorted** [ˋsɔrtɪd] *a.* 安排妥當的 (不用於名詞前)
There are still some problems that need to get sorted before we go on our trip.
在我們去旅行前，還有一些問題需要釐清。

經典旅遊實用句

Have you checked all your [物品]?
你檢查過所有……了嗎？

用途：提醒確認旅行裝備是否準備妥當。

你檢查過所有滑雪裝備了嗎？
Have you checked all your skiing gear?

luggage	行李
hiking gear	健行裝備
camera equipment	攝影器材
winter clothes	冬季衣服

旅人私房筆記　　　　　滑雪相關實用字

ski resort [rɪˈzɔrt]
滑雪勝地

ski goggles [ˈgɑglz]
（滑雪）護目鏡

skiing
滑雪（運動）

ski
vi. 滑雪 & *n.* 滑雪板

ski pole
滑雪杖

Try It Out! 試試看　　請依句意在空格內填入適當的字詞

你可以替我挑選一部電影嗎？我不知道要看什麼。
Can you _____ _____ a movie for me? I have no idea what to watch.

Ans　pick out

旅遊英語輕鬆聊　　參考答案請掃描使用說明頁上 QR code 下載

Do you look cool in ski gear... or like a Michelin Man?

CH 05　Sightseeing · 觀光

185

Unit 45 Taking Pictures
拍照

實境對話 GO！

M Marcus（馬克斯）　　**T** Tori（托莉）

Marcus is on vacation with his wife and asks a passerby, Tori, to take their photo.

M Excuse me, could you take a picture of us, please?

T Of course, it would be my pleasure.

M Thank you. Here's my phone. Please try and get the tower in the **background**.

T No problem. I'll take a **vertical** one. I just press this button here, right?

M Yes, that's right.

T OK. Move back a little and to the left. Stand a little closer together. That's perfect. Three, two, one—say cheese!

M Thanks. Could you take one more, just to be safe?

T Of course. Three, two, one—smile! There you go.

M Thanks for your help.

T You're welcome. Did they **turn out** OK?

M Yes, they look great! The view here is amazing, isn't it? Would you like me to take one of you?

T Sure, thanks. *(Hands Marcus her phone.)*

M Oh, this is the latest model, right? The quality of your photo will be even better than ours!

T Yes, I just got it—don't drop it!

M Haha, don't worry. OK, here we go. Three, two, one! Wow, it looks **awesome**, even if I do say so myself!

T You're right! You should be a **professional** photographer!

M Maybe in another life. Anyway, enjoy the rest of your trip!

T Thanks, you too!

馬克斯和他的妻子正在度假，找了一個叫托莉的路人幫他們拍照。

M 不好意思，請問可以幫我們拍張照嗎？

T 當然，我很樂意。

M 謝謝。這是我的手機。請盡量把後面的塔也拍進去。

T 沒問題。我直著拍。按這個鍵對不對？

M 對。

T 好。稍微後退一點，再往左邊一點。你倆站近一點。這樣很好。三、二、一——笑一個！

M 謝謝。可以再拍一張嗎？比較保險。

T 當然可以。三、二、一——笑一個！好了。

M 謝謝妳幫忙。

T 不客氣。照片拍得還行嗎？

M 行，拍得真好！這裡的景色好好看，不是嗎？要不要我幫妳拍一張？

T 好啊，謝謝。（把她的手機遞給馬克斯。）

M 喔，這是最新款的吧？妳的照片畫質會比我們的更好！

T 對，我剛買的——別掉了！

M 哈哈，別擔心。好，要拍囉。三、二、一！哇，看起來真不錯，我得誇自己一下！

T 對耶！你該當專業攝影師喔！

M 下輩子吧。總之祝妳旅途愉快！

T 謝謝，你們也是！

旅遊字詞補給包

❶ **background** [ˈbækˌɡraʊnd] *n.* 背景
The majestic mountains formed a stunning background for the photograph.
雄偉的山脈為這張照片構成了一幅令人驚嘆的背景。

❷ **vertical** [ˈvɝtɪkl̩] *a.* 垂直的
Please hold your phone in a vertical position to take this photo.
請用手機直著拍這張照片。

❸ **turn out...** 結果是……（常指出乎意料的結果）
Surprisingly, the small party turned out to be a lot of fun.
這個小型派對意外地非常好玩。

❹ **awesome** [ˈɔsəm] *a.* 很棒的
The view from the top of Taipei 101 was absolutely awesome!
從台北 101 頂樓看到的景色真是太棒了！

❺ **professional** [prəˈfɛʃənl̩] *a.* 專業的，職業的 & *n.* 專家；職業選手
Tom decided to hire a professional photographer for his wedding.
湯姆決定為他的婚禮聘請一位專業攝影師。

經典旅遊實用句

Move back a little and [方向]. 往後退一點，再往……一點。

用途：指導姿勢或站位，讓拍照效果更好。

稍微後退一點，再往左邊一點。
Move back a little and to the left.
　　　　　　　　　　　to the right　　　往右一點
　　　　　　　　　　　closer to each other　　互相靠近一點

188

旅人私房筆記 — 拍照姿勢大錦集

Put your hands on your hips.
叉腰。

Cross your arms.
雙手抱胸。

Play with your hair.
玩頭髮。

Look over your shoulder.
回眸。

Look away from the camera.
別過頭。

Give me a big laugh!
燦笑！

Try It Out! 試試看
請依句意在空格內填入適當的字詞

天氣預報說會下雨，但結果卻是個晴朗的好天氣。
The weather forecast predicted rain, but it _____ _____ to be a beautiful sunny day.

> Ans turned out

旅遊英語輕鬆聊
參考答案請掃描使用說明頁上 QR code 下載

If a stranger asks you to take a photo for them, what would you secretly hope?

CH 05
Sightseeing・觀光

Notes

Chapter 06

Food
美食

Unit 46
Enjoying Local Food
品嚐當地美食

Unit 47
Having Afternoon Tea
下午茶

Unit 48
Ordering Food at a Diner
在餐廳點餐

Unit 49
Trying New Foods
嘗試新食物

Unit 50
Tipping Culture
小費文化知多少

Unit 51
Paying the Check
結帳

Unit 46 Enjoying Local Food
品嚐當地美食

實境對話 GO！

🅐 Amy（艾咪）　　🅢 Saul（紹爾）

Saul and Amy are out in the evening.

🅐 We saw a lot of interesting places today.

🅢 Yeah, I'm glad we got to see so much. Now, I'd like to show you another side of Jerusalem.

🅐 You mean it's not just walls and temples?

🅢 Of course not. This is the Arab market in the Old City. I don't think it has changed that much in thousands of years.

🅐 That's so cool. There's lots of colorful art and clothing here.

🅢 You should buy some souvenirs to take home to your friends.

🅐 I definitely will. I also want to try some local snacks.

🅢 Good idea. Before we shop until we drop, let's eat at my favorite restaurant.

🅐 What does the local food taste like?

🅢 You'll love it. It takes the best of all of the cultures staying here. We can start with a pita dish.

🅐 Isn't that bread?

🅢 Yes, but we can add some eggplant, tomatoes, and cucumber to make it taste better.

🅐 What's that brown **stuff** over there?
🅢 It's **hummus**, a special mix of **chickpeas**, garlic, and **olive** oil. You can add some to your **pita bread**.
🅐 Wow, this is delicious!

紹爾和愛咪晚上在外面逛。

🅐 我們今天看了好多有趣的地方。
🅢 對呀，我很高興我們能看這麼多。現在，我要帶妳看耶路撒冷的另一面。
🅐 你的意思是，不只有牆壁和聖殿嗎？
🅢 當然了。這裡是舊城的阿拉伯市場。我想它在這幾千年以來沒有太大的改變。
🅐 好酷。這裡有好多色彩繽紛的藝術品和衣服。
🅢 妳應該買一些紀念品帶回去送給妳的朋友。
🅐 我絕對會的。我也想要嚐嚐當地小吃。
🅢 好主意。在我們買東西買到不行之前，咱們先到我最喜歡的餐廳吃東西吧。
🅐 當地的食物嚐起來怎麼樣？
🅢 妳會愛上它的。這裡的食物集合了當地所有文化的精華。我們可以先來盤皮塔餅。
🅐 那不是麵包嗎？
🅢 是的，但是我們可以加點茄子、番茄和小黃瓜來讓它更美味。
🅐 那邊那個棕色的東西是什麼呀？
🅢 那是鷹嘴豆泥，也就是一種加了鷹嘴豆、大蒜和橄欖油的特殊混和配料。妳可以加一點到妳的皮塔餅裡。
🅐 哇，這好好吃喔！

旅遊字詞補給包

❶ **souvenir** [ˌsuvəˈnɪr] *n.* 紀念品
This seashell is a souvenir from the beach.
這個貝殼是去海灘撿回來的紀念品。

❷ **definitely** [ˈdɛfənɪtlɪ] *adv.* 肯定地，確切地
As Andrew's best friend, Keith will definitely stand up for him.
身為安德魯最好的朋友，凱斯一定會支持他。

❸ **eggplant** [ˈɛgˌplænt] *n.* 茄子
I like adding eggplants to pasta sauce.　我喜歡在義大利麵醬裡放茄子。

❹ **cucumber** [ˈkjukəmbɚ] *n.* 黃瓜
He added a few slices of fresh cucumber to his sandwich.
他在三明治裡加了幾片新鮮小黃瓜。

❺ **stuff** [stʌf] *n.* 東西；事情　→　非正式，常用於事物名稱不詳或內容不重要時
I'm afraid I can't go out with you. I have so much stuff to do this weekend.　我恐怕沒辦法和你出去玩，我這個週末有好多事要做。

❻ **hummus** [ˈhʌməs] *n.* 鷹嘴豆泥
I like to eat vegetable sticks dipped in hummus.
我喜歡用蔬菜棒沾鷹嘴豆泥吃。

❼ **chickpea** [ˈtʃɪkˌpi] *n.* 鷹嘴豆，雞心豆
Mom cooked a pot of hearty chickpea soup today.
媽媽今天煮了一鍋豐盛的鷹嘴豆湯。

❽ **olive** [ˈɑlɪv] *n.* 橄欖
This bottle of olive oil has a very rich flavor.
這瓶橄欖油的味道非常濃郁。

❾ **pita bread** [ˈpitə ˌbrɛd] *n.* 皮塔餅，(扁平且中空的)圓麵包，口袋餅
The pita bread at this restaurant is soft and chewy.
這家餐廳的口袋餅軟而耐嚼。

經典旅遊實用句

What does the local [食物] taste like?
= How does the local [食物] taste?
當地的……嚐起來怎麼樣？

用途：詢問當地食物的口感或特色。

在地食物吃起來如何？
What does the local food taste like?
　　　　　　coffee　　咖啡
　　　　　　dessert　 點心
　　　　　　fruit　　 水果

旅人私房筆記　　耶路撒冷 —— 新與舊的完美結合

猶太民族重要文獻《塔木德》中記載著這麼一段話：「如果上帝給了世界十分美麗，那麼九分給了耶路撒冷，剩下的一分給了世界上的其他地方；如果上帝給了世界十分哀愁，那麼九分給了耶路撒冷，剩下的一分給了世界上的其他人。」

耶路撒冷是一座由多種文化與民族共同交織而成的城市，曾被視為世界的中心，而現已發展為全球前五大快速崛起的創業城市。舊城 (Old City) 區對具有特殊宗教地位的耶路撒冷而言，可說是最特別的地方。該區雖占地僅一平方公里，但重要的宗教聖地均位於此，像是猶太教的西牆、伊斯蘭教的岩石圓頂，還有基督教的聖墓教堂及耶穌的受難之路 —— 苦路，都是極具特殊宗教意義的地點。

The Western Wall 西牆

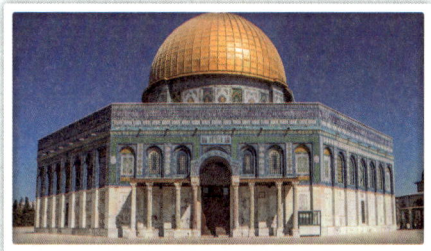

The Dome of the Rock 岩石圓頂

The Church of the Holy Sepulchre 聖墓教堂

Stations of the Cross 苦路

Try It Out! 試試看　　請依句意選出適當的字詞

What's that sticky ＿＿＿＿ on your shirt?
Ⓐ souvenir　　Ⓑ staff　　Ⓒ case　　Ⓓ stuff

Ans Ⓓ

旅遊英語輕鬆聊　　參考答案請掃描使用說明頁上 QR code 下載

Would you rather try a brand-new local dish or stick to something familiar when traveling?

Unit 47 Having Afternoon Tea
下午茶

實境對話 GO！

K Kate（凱特） **A** Annie（安妮）

Kate and her friend Annie are having afternoon tea.

K So, what is this we're having, Annie?

A We're having afternoon tea, like I'd have back when I was in England.

K What's the history?

A Once there was a hungry **duchess**. She couldn't wait for dinner, so she created afternoon tea.

K Being a duchess must be nice. Besides tea, what are we going to have?

A Look, it's coming now. See how there are three **levels** of plates? Each one is its own **course**.

K Where do we start? That cake looks delicious.

A We start from the bottom with the sandwiches. Then we move up to the middle **scones**, and finally reach the top sweets. How do you take your tea, dear?

K One sugar and a bit of cream, please.

A Watch your hand. We must start with the sandwiches.

K Why does it matter?

A We English value order, especially during our meals. Try it this way, and if you don't like it, next time we'll go to an American brunch.

K Sounds good to me.

凱特和她朋友安妮正在享用下午茶。

K 那麼，我們現在吃的是什麼呢，安妮？

A 我們在吃下午茶，就像以前我待在英國時那樣。

K 下午茶的由來是什麼呢？

A 據說以前有位公爵夫人餓了。她等不及吃晚餐，所以就發明了下午茶。

K 身為公爵夫人的感覺一定很棒。除了茶，我們還會吃什麼？

A 妳看，這就來了。妳有注意到為什麼盤子要分三層放了嗎？每層的點心都不同。

K 我們要從哪層開始吃？那蛋糕看起來很好吃。

A 我們從最底層的三明治開始吃。然後再吃中間的司康，最後再吃最上層的甜點。妳的茶要不要加糖或奶精呢，親愛的？

K 請給我一匙的糖和一點奶精。

A 管好妳的手唷。我們必須從三明治開始吃。

K 這很重要嗎？

A 我們英國人很重視順序，尤其是用餐時。試著照這個順序吃，如果妳不喜歡，下次我們去吃美式早午餐。

K 聽起來不錯。

旅遊字詞補給包

❶ **duchess** [ˈdʌtʃɪs] *n.* 公爵夫人；女公爵
The duchess gave the crowd a kind smile.
公爵夫人向群眾露出親切的微笑。

❷ **level** [ˈlɛvḷ] *n.* 層；程度，水準
Students at this level may have problems expressing themselves clearly in English.
這個程度的學生可能還很難用英語清楚表達自己的想法。

❸ **course** [kɔrs] *n.* 一道菜；課程；過程
Alice enjoyed the delicious dessert even more than the main course.
愛麗絲喜歡這份美味的甜點甚於主菜。

❹ **scone** [skon] *n.* 司康 (一種麵點)
We prepared scones, sandwiches, and tea for afternoon tea.
我們為下午茶準備了司康、三明治和茶。

❺ **matter** [ˈmætɚ] *vi.* 重要，要緊，有關係
As long as you're healthy, that's all that matters.
只要你健康，那就是最重要的事情。

❻ **value** [ˈvælju] *vt.* 重視；珍惜
We all value Mr. Nelson's advice because he is an expert in this field.
我們都很重視尼爾森先生的意見，因為他是這一行的專家。

經典旅遊實用句

We must start with [東西].　　我們必須從……開始吃。

用途：說明正確的流程或順序，尤其適合正式場合說明。

我們要從三明治開始吃。
We must start with the sandwiches.
　　　　　　　 the salad　　　沙拉
　　　　　　　 the appetizers　開胃菜
　　　　　　　 the soup　　　 湯

旅人私房筆記 　　**下午茶順序**

AFTERNOON TEA ORDER

Start from the bottom (salty) to the top (sweet).
從下層（鹹點）開始往上層（甜點）吃。

Top（上層）：
the sweets
甜點
fruit tarts
水果塔
cakes
蛋糕

Three levels of plates
三層盤

Middle（中層）：
scones
司康

Bottom（底層）：
sandwiches
三明治

Try It Out! 試試看　請依句意在空格內填入適當的字詞

❶ 你穿什麼都不要緊，只要人來參加派對就好。
It doesn't _____ what you wear as long as you come to the party.

❷ 飲料和前菜通常都是在主菜前上菜。
Drinks and side dishes are usually served before the main _____.

Ans ❶ matter　❷ course

旅遊英語輕鬆聊　參考答案請掃描使用說明頁上 QR code 下載

Do you usually follow the rules of food "order", or do you just eat whatever you want first?

CH 06　Food・美食

Unit 48 Ordering Food at a Diner
在餐廳點餐

實境對話 GO!

W Waitress（女服務生）　　**N** Nicholai（尼可萊）

Nicholai enters a restaurant, takes a seat, and begins looking through the menu. A waitress approaches him.

W Hello, Sir. Do you know what you'd like to order yet?

N Actually, I've never been here before. Are there any dishes that you suggest I try?

W Well, everything on our menu is sure to satisfy you. Personally, I can't get enough of our chicken sandwich.

N That sounds good. Is it your most popular dish?

W Oh, no. Not even close. Most of the regulars get the buffalo wings or one of the BBQ platters.

N Well, I'm quite hungry, so I'll try some buffalo wings first. Then I'll have the chicken sandwich that you mentioned for the main course.

W Sure thing. Would you like fries or steamed vegetables with the chicken sandwich?

N A side of vegetables would be nice. Oh, and another thing; please hold the mayonnaise on the chicken sandwich.

W Oh, don't worry. We don't use mayonnaise in our chicken sandwiches.

N Wonderful.

W Would you like something to drink? We have an excellent **selection** of tea and coffee.

N I'll have a sweet iced tea to drink.

W No problem. I'll get your order in and bring you that drink right away.

尼可萊走進一家餐廳，找了座位坐下，並開始看菜單。一位女服務生走向他。

W 哈囉，先生。您決定好想點什麼了嗎？

N 其實，我沒有來過這裡。妳會建議我嘗試哪些菜呢？

W 嗯，我們菜單上的每道菜都保證讓您滿意。就我個人來說，我非常喜歡我們的雞肉三明治。

N 那聽起來不錯。那是你們最熱銷的菜嗎？

W 喔，不。差得遠了。大部分的常客都會點水牛城辣雞翅或是其中一道燒烤拼盤。

N 嗯，我還蠻餓的，所以我會先吃吃看水牛城辣雞翅。然後我要來份妳提到的雞肉三明治作為主菜。

W 好的。您想要薯條還是蒸蔬菜來搭配雞肉三明治嗎？

N 蔬菜附餐聽起來不錯。喔，還有一件事；請不要在雞肉三明治上加美乃滋。

W 喔，不用擔心。我們的雞肉三明治是不使用美乃滋的。

N 太好了。

W 您想喝點什麼嗎？我們有一些很棒的茶和咖啡任君選擇。

N 我要來一杯甜的冰茶。

W 沒問題。我會把您的訂單送進廚房並立刻送上您的飲料。

旅遊字詞補給包

❶ diner [ˈdaɪnɚ] *n.* 餐館
This diner is famous for its hearty burgers and milkshakes.
這家餐館以漢堡料爆多和奶昔超濃聞名。

❷ look through... 瀏覽……，快速地看……
Elias looked through the newspaper for interesting stories.
伊萊亞斯瀏覽報紙想找些有趣的報導。

CH 06

Food・美食

❸ **regular** [ˈrɛgjələ] *n.* 常客
Mr. Li is a regular at this café; he comes almost every day.
李先生是這間咖啡廳的常客，幾乎每天都來。

❹ **buffalo wings** [ˈbʌfəˌlo ˈwɪŋz] *n.* 水牛城辣雞翅（多用複數）
We ordered a serving of spicy buffalo wings as an appetizer.
我們點了一份水牛城辣雞翅當開胃菜。

❺ **platter** [ˈplætə] *n.* 拼盤；圓盤，淺盤
This fruit platter looks both fresh and colorful.
這個水果拼盤看起來很新鮮，還有好幾種顏色。

❻ **steamed** [stimd] *a.* 蒸的（過去分詞作形容詞用）
Annie loves steamed buns because they're hot and delicious.
安妮喜歡蒸過的包子，因為它們又熱又美味。

❼ **side** [saɪd] *n.* 附餐
I ordered a hamburger, and I chose fries as my side.
我點了漢堡，附餐選薯條。

❽ **mayonnaise** [ˈmeəˌnez] *n.* 美乃滋
This sandwich has a thick layer of mayonnaise on it.
這個三明治上塗了厚厚一層美乃滋。

❾ **selection** [səˈlɛkʃən] *n.* 可供挑選的東西；選擇
This library has a wide selection of scientific journals for readers.
這間圖書館有豐富的科學期刊供讀者挑選。

經典旅遊實用句

Please hold the [配料].　　請不要加……配料。

用途：請求不要加某個配料，在餐廳點餐十分實用。

請不要在雞肉三明治上加美乃滋。
Please hold the mayonnaise on the chicken sandwich.
　　　　　　　onions　　洋蔥
　　　　　　　cheese　　起司

旅人私房筆記　　　　　認識菜單寫什麼

MENU

APPETIZERS
開胃菜

SALADS
沙拉

MAIN DISHES
主菜

SNACKS
點心

SEAFOOD
海鮮

DRINKS
飲料

CH 06

Food · 美食

Try It Out! 試試看　　請依句意在空格內填入適當的字詞

記得試吃這家店所提供種類繁多的冰淇淋。
Remember to sample the wide _____ of ice cream this store offers.

Ans　selection

旅遊英語輕鬆聊　　參考答案請掃描使用說明頁上 QR code 下載

When you go to a restaurant for the first time, do you usually play it safe or try something adventurous?

203

Unit 49 Trying New Foods
嘗試新食物

實境對話 GO！

W Waiter（服務生）　　**M** Maggie（瑪姬）

Maggie is in a restaurant, talking to the waiter about her food.

W Good evening, ma'am. Is everything OK with your meal?

M The steak is cooked to **perfection**—it's juicy and **tender**.

W I'm glad to hear that.

M But, uh…

W Is there a problem with another part of your meal?

M The blue cheese **sauce** is a little strong for me. I'm struggling with the flavor—it's kind of **intense**.

W That's understandable if you're not used to blue cheese. It's made with a special **mold** that helps the cheese age and creates a strong, powerful flavor. It's definitely an **acquired** taste.

M I think it may take me a bit longer to acquire it! To be honest, I didn't quite realize what I was ordering.

W At least you got it on the side and not on the steak. Would you like me to bring you a different sauce to try?

M Which one do you recommend?

W How about peppercorn sauce? It's creamy and mildly spicy. It's still somewhat bold, but it's not overpowering like the blue cheese sauce.

M That sounds more suitable for my tastes. Thank you for your help.

W You're welcome, ma'am. I'll be right back with your peppercorn sauce.

餐廳裡，瑪姬正和服務生討論她的餐點。

W 晚安，女士。您的餐點還滿意嗎？

M 牛排煎得恰到好處 —— 不但多汁還很嫩。

W 那真是太好了。

M 不過呢……

W 餐點的其他部分有問題嗎？

M 藍紋起司醬對我來說口味有點太重。我有點吃不慣 —— 味道過於刺激了。

W 不習慣藍紋起司的味道是可以理解的。它裡面有種特殊的黴菌，所以味道特別濃烈。的確是要慢慢習慣它。

M 我應該會需要很多時間來習慣吧！說實話，我點的時候不知道是這種味道。

W 至少您的醬是分開放的，沒有直接淋在牛排上。要不要我幫您換一種醬料試試？

M 你有推薦的嗎？

W 胡椒醬怎麼樣？它比較濃，有一點點辣。口味仍然是重的，但沒到藍紋起司醬那樣強烈。

M 感覺比較合我的口味。謝謝你的幫忙。

W 不客氣，女士。我馬上就把胡椒醬拿過來。

旅遊字詞補給包

❶ **perfection** [pɚˋfɛkʃən] *n.* 完美
The baker aimed for perfection with his cakes.
這位烘培師傅力求烘出完美的蛋糕。

CH 06

Food・美食

205

❷ **tender** [ˈtɛndɚ] *a.* (肉或蔬菜)嫩的，軟的；溫柔的；(年紀)稚嫩的
The steak was so tender that it almost melted in my mouth.
這塊牛排非常軟嫩，幾乎入口即化。

❸ **sauce** [sɔs] *n.* 醬汁
Mandy made a special sauce using fresh herbs from her garden.
曼蒂用花園裡現摘的香料草做了一種特別的醬。

❹ **intense** [ɪnˈtɛns] *a.* 強烈的
The sunlight on the beach was intense. 海灘上的陽光很強烈。

❺ **mold** [mold] *n.* 黴菌
There was green mold growing on the old bread.
那塊擺太久的麵包上長了綠色黴菌。

❻ **acquired** [əˈkwaɪrd] *a.* 養成的，習得的
Megan thinks that black coffee is an acquired taste.
梅根覺得黑咖啡需要時間來習慣。

❼ **mildly** [ˈmaɪldlɪ] *adv.* 略微地；溫和地
This dish is mildly spicy, and most people can accept it.
這道菜的味道微辣，大部分人都能接受。

❽ **suitable** [ˈsutəbḷ] *a.* 適當的
be suitable for... 適合……
This type of soil is suitable for growing vegetables.
這種土壤適合種蔬菜。

經典旅遊實用句

Is everything OK with your [餐點]? 您的……還滿意嗎？

用途：餐廳服務生常見的關心問候。
您的餐點還滿意嗎？
Is everything OK with your meal?

dish	菜餚
order	點的餐
pasta	義大利麵

旅人私房筆記　　各式牛排醬料英文怎麼說？

blue cheese sauce 藍紋起司醬
鹹中帶酸，甚至帶一點刺鼻感或發酵味

peppercorn sauce / black pepper sauce 黑胡椒醬
帶點辣味但溫和

mushroom sauce 蘑菇醬
濃郁香氣，口感順滑

mustard sauce 芥末醬
有點嗆辣但提味效果好

Try It Out! 試試看　　請依句意在空格內填入適當的字詞

這裡的氣候一年四季都適合戶外活動。
The climate here is ＿＿＿＿＿ ＿＿＿＿＿ outdoor activities all year round.

Ans suitable for

旅遊英語輕鬆聊　　參考答案請掃描使用說明頁上 QR code 下載

Have you ever ordered something at a restaurant and wished you hadn't?

CH 06　Food・美食

Unit 50 Tipping Culture
小費文化知多少

實境對話 GO！

A Alex（艾力克斯）　　**S** Sandra（珊卓）

Alex and Sandra are tourists visiting the US. They are about to pay their bill at a restaurant.

A That meal was great. How much do we pay?

S Let me see the bill. Our meal itself was $45.50. After adding the tax, the total amount comes to $49.83. Should we just pay that?

A America has a tipping culture, you know. It would be rude if we didn't tip.

S Oh, right. I've been in Japan for too long.

A Do they not tip in Japan?

S No, they don't. Actually, if you try to tip, it is seen as rude. The worker will likely not accept it. It's not part of their culture at all.

A In America, it's the opposite. Many different professions rely on tipping, such as taxi drivers and hairdressers.

S It kind of makes everything more expensive, though, doesn't it?

A It does. But usually the quality of service is better because tips are based on workers' performances. Especially in restaurants, waiters and waitresses aren't paid very well by the hour, so a big part of their **salaries** comes from tips.

🅂 How much should we tip our waiter today? I think he did pretty well, and he was very friendly, too.

🄰 For good service, I usually give 20% of the **original** bill, without the tax included.

🅂 That means we should give him $9.10. Why don't we just **round up to** $10?

🄰 Sure. That works.

艾力克斯和珊卓是到訪美國的遊客。他們正準備在一間餐廳結帳。

🄰 那頓飯真不錯。我們要付多少錢？

🅂 讓我看一下帳單。我們的餐點本身是 45.50 美元。加稅金後，總金額是 49.83 美元。我們只要付這樣就好嗎？

🄰 美國有小費文化，妳知道的。如果我們不給小費的話會很失禮。

🅂 喔，對耶。我在日本待太久了。

🄰 在日本不給小費的嗎？

🅂 對，不給。事實上，如果你試著要給，會被視為很失禮的舉動。員工很可能不會接受小費。這完全不是他們文化的一部分。

🄰 在美國情況恰好相反。很多不同的職業都仰賴小費，例如計程車司機和髮型師。

🅂 但這會讓所有東西都變得更貴，不是嗎？

🄰 是啊。但通常服務的品質會比較好，因為小費是根據員工的表現給的。特別是在餐廳，服務生的時薪並不是很高，所以他們有一大部分的薪水是來自小費。

🅂 我們該給今天的服務生多少小費？我覺得他做得非常好，而且他也很友善。

🄰 對於良好的服務，我通常會給原本帳單上的百分之二十，不含稅金。

🅂 那就是我們應該給他 9.10 美元。我們何不乾脆取整數給他十美元？

🄰 當然。那行得通。

旅遊字詞補給包

三態為：tip, tipped, tipped

❶ tip [tɪp] *vi.* 給小費 & *vt.* 給……小費 & *n.* 小費
In American restaurants, tipping is very common.
在美國餐廳裡，給小費是非常普遍的事。
Betty tipped the waiter generously because of his good service.
這名服務生服務很好，所以貝蒂給了他很多小費。
Calvin gave the taxi driver US$1 as a tip.
卡爾文給了那名計程車司機一塊美金作為小費。

❷ be about to V　　正要 / 即將要……
Rumor has it that the bank is about to close down.
謠傳該銀行即將歇業。

❸ profession [prəˋfɛʃən] *n.* 職業
Lucy wants to make teaching her profession.
露西想以教書為業。

❹ rely on...　　依賴 / 依靠……
Many firms rely on the internet to do their business.
許多公司都仰賴網路處理業務。

❺ performance [pɚˋfɔrməns] *n.* 表現
Peter's performance was worth praising.
彼得的表現值得讚揚。

❻ salary [ˋsælərɪ] *n.* 薪水
Darren has difficulty living on such a small salary.
達倫靠這麼微薄的薪水難以維生。

❼ original [əˋrɪdʒənḷ] *a.* 原先的，最初的
The old car still has many of its original parts.
這輛老爺車許多原始的零件都還在。

❽ round up to...　　將……(數字、價格等) 無條件進位
The teacher told the students to round up to the nearest number.
該老師告訴學生要將數字無條件進位。

經典旅遊實用句

How much do we pay? 我們要付多少錢？

用途：詢問總金額，結帳時常用。

我們要付多少錢？
How much do we pay?
　　　　 is the total　　總金額是多少？
　　　　 do we owe　　 我們要付多少錢？

旅人私房筆記

小費給還是不給？真是個難題！

各國文化大不同，以下就來帶你一窺各國小費文化：

一定要給小費的國家
- 在美國、加拿大及葡萄牙等國家是一定要給小費的，小費金額從整體消費的 10% 到 25% 都有。

不需給小費的國家
- 像義大利、日本和瑞士等國家則不需另外給小費，因為小費通常已列入帳單中，若給服務生小費，很可能會被誤會是歧視、瞧不起對方。

自願性給小費的國家
- 澳洲、泰國及比利時等國家是可給可不給，小費行為算是罕見，但客人如果想給，對方通常也會心存感激地接受。

Try It Out! 試試看

請從右欄選出最適合的句子來完成該句

1. Jerry tipped the man US$10
2. Ted was about to leave
3. Emily relies on her father to help her

A. every time she gets into trouble.
B. for carrying his luggage to his room.
C. when Dora showed up.

Ans　❶ B　　❷ C　　❸ A

旅遊英語輕鬆聊

參考答案請掃描使用說明頁上 QR code 下載

Do you think tipping makes service better? Why or why not?

CH 06
Food・美食

211

Unit 51 Paying the Check
結帳

實境對話 GO !

M Melvin（馬文）　　**L** Leticia（萊蒂西亞）

Melvin and his friends have just finished their meal in a restaurant, and he is asking Leticia, a waitress, for the check.

M Excuse me, could we get the check, please?

L Of course. Will you be paying together or separately?

M We'll pay all together.

L OK. I'll be right back. *(Goes to get the check.)* Here you go.

M Thanks. Hmm… There seems to be a problem. We ordered two desserts—not four.

L Oh, I'm sorry about that. Let me just double-check our order records. You're right. They were added by mistake. I've taken them off the bill. I'm very sorry about the confusion.

M No problem. Thanks for sorting that out. Is service included?

L Yes, a 10% service charge has been added. Feel free to tip extra if you'd like. Would you like to pay by cash or card?

M We'll pay by card.

L No problem. Here's the machine. As you've spent over 50 euros, you'll have to **insert** your card into the machine and enter your PIN.

M OK... Done. Thanks for everything—the food was excellent.

L You're welcome. Here's your **receipt**. Enjoy the rest of your evening!

M Thank you. Have a great evening, too!

馬文和他的朋友們剛在餐廳用完餐，他正向女服務生萊蒂西亞要帳單。

M 不好意思，可以幫我們結帳嗎？

L 好的，請問你們要一起付還是分開付呢？

M 一起付。

L 好的，馬上回來。（去拿帳單。）這是你們的帳單。

M 謝謝。嗯……好像有點問題。我們點了兩份甜點，不是四份。

L 喔，真抱歉。我馬上回去核對一下我們的點餐紀錄。您是對的，那是誤加的。我已經從帳單上移除了。非常抱歉造成您的困擾。

M 沒關係，謝謝妳幫忙處理。請問有包含服務費嗎？

L 有的，已經加了 10% 的服務費。如果您願意，也可以額外給小費。請問您要用現金還是信用卡付款呢？

M 用信用卡。

L 沒問題。這是刷卡機。由於您消費超過 50 歐元，您需要將卡片插入機器並輸入 PIN 碼。

M 好的……好了。謝謝你們，食物很棒。

L 不客氣。這是您的收據。祝您有個愉快的夜晚！

M 謝謝。也祝您有個美好的夜晚！

旅遊字詞補給包

❶ by mistake　　弄錯
We were given the wrong dish by mistake.
我們的菜送錯了。

❷ **confusion** [kənˈfjuʒən] *n.* 混亂；混淆，困惑
The unclear sign caused confusion.
不明確的標誌讓大家看得一頭霧水。

❸ **extra** [ˈɛkstrə] *adv.* 額外地 & *a.* 額外的 & *n.* 加購品
Drive extra carefully in this bad weather.
在這種惡劣的天氣裡，開車要格外小心。

❹ **insert** [ɪnˈsɝt] *vt.* 插入
George carefully inserted the memory card into the camera.
喬治小心地將記憶卡插入相機。

❺ **receipt** [rɪˈsit] *n.* 收據，發票
Dolly checked the receipt to make sure everything was correct.
朵莉核對了收據，以確保所有項目都正確。

經典旅遊實用句

Excuse me, could we get the..., please?
不好意思，可以給我們……嗎？

用途：用餐完畢後，禮貌地要求結帳。

不好意思，可以給我們帳單嗎？
Excuse me, could we get the check, please?

| bill | 帳單 |
| receipt | 收據（通常是付完錢後說的）|

旅人私房筆記　　餐廳結帳方式百百種

我們一起來看一下在餐廳有哪些常見的結帳方式吧！

Paying in cash　現金付款
- 在日本、臺灣、泰國等地，現金仍非常普遍，即使餐廳可以刷卡，很多人也偏好付現。
- 在歐洲一些小餐館或市集 (特別是德國、義大利)，小店常常只收現金。

Paying by credit card　信用卡付款
- 美國：刷卡後通常只需簽名，不用輸入 PIN 碼。
- 歐洲（像是法國、德國）：大多數情況需要插卡並輸入 PIN 碼。

Mobile payment　行動支付
- 在北歐國家（如瑞典、挪威），行動支付幾乎取代了現金。
- 中國大陸大多使用微信支付、支付寶，美國的 Apple Pay 等也越來越常見，但在部分小餐館仍偏好現金或刷卡。

Splitting the bill　分開結帳
- 在美國、澳洲、加拿大，AA 制（各付各的）非常普遍，服務生也習慣幫客人分帳。
- 在南歐（像西班牙、義大利），朋友聚餐時通常會一人付全部，然後私下平分，較少在餐廳分開結帳。
- 有些餐廳（特別是繁忙時段）不願意幫忙分帳，會貼出告示提醒。

Paying at the counter　吃完到櫃檯結帳
- 在日本、韓國，通常吃完後要自己拿帳單去櫃檯付款。
- 在西方國家（如美國、英國），餐廳一般是服務生直接拿帳單到桌邊結帳，客人不需要移動。

Try It Out! 試試看　請依句意在空格內填入適當的字詞

我不小心拿錯你的雨傘；我以為是我的。
I took your umbrella _____ _____; I thought it was mine.

Ans by mistake

旅遊英語輕鬆聊　參考答案請掃描使用說明頁上 QR code 下載

Do you prefer paying with cash or card when you eat out? Why?

Notes

Chapter 07 Shopping
購物

Unit 52
Exchanging Currency
兌換外幣

Unit 53
Trying on Clothes
試穿衣服

Unit 54
Bargaining in a Store
殺價

Unit 55
Requesting a Return or Refund
要求退貨退款

Unit 56
Duty-Free Shopping at the Airport
免稅店購物樂

Unit 57
Getting a Tax Refund
申請退稅

25% OFF

217

Unit 52 Exchanging Currency
兌換外幣

實境對話 GO！

E Erin（艾琳）　　**R** Ryan（萊恩）

Ryan is helping Erin exchange some currency at the foreign exchange counter at the airport.

E Hi. I have a couple of different currencies that I'd like to exchange for New Taiwan dollars.

R Of course. How many different currencies do you have?

E I'm actually not sure. I've been traveling these past few months and have collected quite a few. **Off the top of my head**, I have money from Japan, Korea, Cambodia, and Thailand.

R Sure. How about you put everything you want to exchange on the counter? I'll **count** it all **out** for you and tell you how much you'll get back in New Taiwan dollars.

E That sounds good.

R Great. I'll check these for you using today's exchange **rate**. First, your ¥6,000 will **amount to** NT$1,731.

E I think I have some more yen but it's all in coins. Can I also exchange those?

R I'm sorry, but we only accept bills, not coins.

E Oh, no. In that case, I'm probably going to end up with a lot of **loose change**.

R Unfortunately, I don't think you will be able to get those exchanged anywhere. Perhaps you can just hold onto the change and keep it as a souvenir from the countries you visited.

E I guess that's one way to use it.

R Just give me a few minutes while I count out the other currencies for you… OK. Your Korean won can be exchanged for NT$630, the Cambodian riel amounts to NT$1,550, and you'll get NT$370 for your Thai baht.

E Great. Thanks for your help!

萊恩正在機場的兌換外幣櫃檯幫艾琳兌換一些外幣。

E 嗨。我有一些不同的貨幣想要換成新臺幣。

R 沒問題。請問您有幾種不同貨幣？

E 我其實不確定。我過去幾個月都一直在旅行而且已經積了不少。就我現在想得到的，我有日本、韓國、柬埔寨以及泰國的貨幣。

R 好的。那請您將所有想兌換的貨幣放在櫃檯上好嗎？我將為您全部點清再告知您能拿回多少新臺幣。

E 聽起來不錯。

R 很好。我將以今天的匯率替您核算這些外幣。首先，您的六千日圓總共能換新臺幣一千七百三十一元。

E 我想我還有一些日圓，但都是硬幣。我也能兌換那些嗎？

R 很抱歉，但我們只收鈔票不收硬幣。

E 喔，不。那樣的話，我可能最後會有一堆零錢。

R 很不幸的是，我不認為您將有辦法在任何地方兌換那些零錢。也許您可以就保留著它們，當作去過的國家的紀念品。

E 我想也是可以這樣做。

R 請給我幾分鐘替您點清其他貨幣……好了。您的韓元能換新臺幣六百三十元，柬埔寨瑞爾總共可換新臺幣一千五百五十元，然後您的泰銖能換新臺幣三百七十元。

E 好的。感謝妳的幫助！

旅遊字詞補給包

rate of exchange 匯率

❶ **exchange** [ɪksˋtʃendʒ] *vt. & n.* 交換；更換
Prices are always changing according to the rate of exchange.
價格總是隨匯率而變動。

❷ **off the top of sb's head**　某人不假思索地；當下浮現某人腦海地
I cannot tell you her exact address off the top of my head. But I can check that for you.
我現在無法馬上想到並跟您說她確切的住址。但我可以幫您確認。

❸ **count sth out / count out sth**　逐一數清某物
Please wait while I count these coins out for you.
請稍後，我來為您逐一點清這些硬幣。

❹ **rate** [ret] *n.* (費用，價格的) 比率
Martha is rich, so she needs to pay tax at a very high rate.
瑪莎很有錢，所以她得以很高的稅率繳稅。

❺ **amount** [əˋmaʊnt] **to...**　總計為……；達到……
John's annual income amounts to one million dollars.
約翰的年收入達百萬美元。
annual [ˋænjʊəl] *a.* 每年的

❻ **loose change** [ˏlus ˋtʃendʒ] *n.* (包包或口袋裡的) 零錢
Can you pay? I don't have any loose change.
你可以付錢嗎？我沒有任何零錢。

經典旅遊實用句

I have a couple of different [貨幣] that I'd like to exchange.　我有幾種不同的……想要兌換。

用途：表明自己要兌換多種外幣。

我有幾種不同的貨幣想要兌換。
I have a couple of different currencies that I'd like to exchange.
bill　紙鈔

旅人私房筆記 — 認識常見貨幣

New Taiwan dollar 新臺幣

won [wən] 韓元

yen [jɛn] 元（符號：¥）

RMB (Renminbi) 人民幣

baht [bɑt] 泰銖

United States dollar 美金

Try It Out! 試試看
請依句意在空格內填入適當的字詞

你有零錢可以投販賣機嗎？
Do you have any _____ _____ for the vending machine?

Ans loose change

旅遊英語輕鬆聊
參考答案請掃描使用說明頁上 QR code 下載

If you could only keep one kind of foreign coin as a souvenir, which one would you choose and why?

Unit 53 Trying on Clothes
試穿衣服

實境對話 GO！

J Juan（璜）　**M** Mia（蜜亞）

Mia is looking at clothes in a store and is **approached** by Juan, a sales **assistant**.

J Hi. Can I help you with anything?

M I was just looking at this dress. Do you have it in a medium?

J Let me check… Yes, here you go.

M Can I try it on?

J Of course. The fitting rooms are just over there, next to the **cashier**.

M Great, thanks. *(Goes to try on the dress.)*

J So, what do you think?

M It fits well, and I like the style.

J Yes, the style is on-trend this summer.

M I'm not sure the color suits me, though. Does it come in any other colors?

J Yes—we have it in light blue and pink as well.

M I'll try the blue, please.

J Let me grab that one for you… Here you go.

M Thanks. *(Goes to try on the next dress.)*

J Wow! That looks great on you!

M Thanks—I agree! I'll take it!

J Perfect. Are you looking for anything else today?

M I also want to check out the denim jackets.

J They're at the back of the store, on the right. Would you like me to hold onto your dress while you browse the jackets?

M That would be great, thanks.

蜜亞在一家店裡看衣服，店員璜走近她。

J 嗨，請問有什麼需要我幫忙的嗎？

M 我正在看這件洋裝。請問有中號嗎？

J 讓我看看……，有的，這件給您。

M 我可以試穿嗎？

J 當然可以。試衣間在那邊，收銀臺旁邊。

M 太好了，謝謝。（去試穿洋裝。）

J 覺得如何？

M 很合身，而且我喜歡這個款式。

J 對呀，今夏流行這款。

M 不過，我不確定這個顏色是否適合我。請問還有其他顏色嗎？

J 有的，我們還有淺藍色和粉紅色。

M 請給我藍色的試試。

J 我幫您拿那件……，給您。

M 謝謝。（去試穿下一件洋裝。）

J 哇！這件您穿起來真好看！

M 謝謝 —— 我也覺得！我買了！

J 太棒了。您還想看看其他東西嗎？

M 我也想看看牛仔外套。

J 牛仔外套在後面，右邊。您逛外套的時候，需要我幫您保留這件洋裝嗎？

M 那太好了，謝謝。

旅遊字詞補給包

❶ **approach** [əˋprotʃ] *vt.* & *vi.* 接近 & *n.* 方法
The cat slowly approached the sleeping mouse.
那隻貓躡手躡腳地靠近正在睡覺的老鼠。

❷ **assistant** [əˋsɪstənt] *n.* 店員；助理 & *a.* 助理的
The sales assistant helped me find the right-sized sneakers.
店員幫我找到了合適尺寸的球鞋。

❸ **cashier** [kæˋʃɪr] *n.* (商店、銀行等的) 收銀員；出納員
The cashier gave me the receipt and my change.
收銀員給了我收據和零錢。

❹ **denim** [ˋdɛnɪm] *n.* 牛仔布 (不可數)
Cindy loves wearing denim jackets in the spring.
春天時辛蒂喜歡穿牛仔外套。

❺ **browse** [braʊz] *vi.* & *vt.* & *n.* 瀏覽，隨意觀看
Gary spent the afternoon browsing in the art gallery.
蓋瑞花一個下午在美術館裡逛逛看看。

經典旅遊實用句

The fitting rooms are just [位置].　　試衣間在……。

用途：指引試衣間位置。

試衣間在那邊。
The fitting rooms are just over there.
　　　　　　　　　　by the window　　在窗戶旁邊
　　　　　　　　　　next to the mirror　　在鏡子旁邊

旅人私房筆記 — 從頭到腳，服飾英文輕鬆記！

上衣類 Tops
- **shirt** 襯衫
- **t-shirt** T恤
- **blouse** 女用襯衫 / 上衣
- **sweater** 毛衣
- **hoodie** 連帽上衣
- **tank top** 無袖上衣 / 背心

下身類 Bottoms
- **pants** 長褲
- **jeans** 牛仔褲
- **shorts** 短褲
- **skirt** 裙子
- **leggings** 緊身褲

外套類 Outerwear
- **jacket** 夾克 / 外套
- **coat** 較厚較長的外套
- **cardigan** 開襟針織衫

洋裝類 Dresses
- **dress** 洋裝
- **evening gown** 晚禮服
- **sundress** 夏日輕便連身裙

鞋子類 Shoes
- **sneakers** 運動鞋
- **sandals** 涼鞋
- **boots** 靴子
- **heels** 高跟鞋

配件類 Accessories
- **scarf** 圍巾
- **hat** 帽子
- **gloves** 手套
- **belt** 皮帶

Try It Out! 試試看 　請依句意在空格內填入適當的字詞

一隻友善的狗搖著尾巴向我們走近。
A friendly dog ＿＿＿＿＿＿ us, wagging its tail.

Ans approached

旅遊英語輕鬆聊 　參考答案請掃描使用說明頁上 QR code 下載

What's more important to you when shopping: style, comfort, or price?

CH 07　Shopping · 購物

Unit 54 Bargaining in a Store
殺價

實境對話 GO！

A Albert（亞伯特）　　**R** Rosa（羅莎）

> Albert is browsing in a **souvenir** store and talking to Rosa, who works there.

A This elephant woodcarving is beautiful.

R Thank you. Yes, it is. It was hand-carved by a local **craftsman**.

A My mom would love it—she is fascinated by elephants. How much is it?

R It's only $50.

A That's a bit more than I was hoping to spend. Is there any way you could make it cheaper?

R Hmm… I could offer it to you for $45.

A $45 is still a bit too high. I'm not sure I can **afford** that. Can you give me a bigger **discount**? I think $35 is a fair price. I saw some similar ones for that price at another store.

R I can **guarantee** they aren't as high-quality as the ones we sell. I'm sorry, but I can't sell it for $35. $42 is my best price.

A Could you throw in something else to sweeten the deal?

R I can give you this model of the town's famous church. It's a great souvenir for you or a loved one.

A OK, then. But I'll give you $40 for both. **Take it or leave it**.

R I'll take it—it's a deal!

A Thank you.

R Thank you. I hope you enjoy your purchase.

> 亞伯特正在逛紀念品店，跟那裡的工作人員羅莎交談中。

A 這木雕大象很漂亮。

R 謝謝。沒錯，這是在地工匠手工雕刻的。

A 我媽一定會喜歡 —— 她是象迷。這個多少錢？

R 只要五十美元。

A 比我的預算貴一點。能算便宜一點嗎？

R 嗯⋯⋯我可以算你四十五。

A 四十五還是有點貴，我買不下手。可以多給點折扣嗎？我覺得三十五比較合理。我在別家店看到類似的東西是賣這個價錢。

R 我敢保證那家的品質絕對不如我們家。很抱歉，三十五沒辦法。四十二最低價給你。

A 妳送點什麼給我吧，這樣比較划算。

R 我可以送你這個鎮上知名教堂的小模型。送禮自用都很好。

A 好吧。那我這兩樣給妳四十。不行就算了。

R 好吧 —— 成交！

A 謝謝。

R 謝謝你。希望你喜歡買到的東西。

旅遊字詞補給包

❶ **bargain** [ˈbɑrgən] *vi.* 講價，討價還價 & *n.* 很划算的東西（可數）
Adam tried to bargain with the street vendor for a lower price.
亞當和路邊攤販殺價，希望能買到更低的價格。

❷ **souvenir** [ˈsuvəˌnɪr] *n.* 紀念品；紀念物
Bess collected souvenirs from every country she visited.
貝絲每去一個國家都要收集那裡的紀念品。

❸ **craftsman** [ˈkræftsmən] *n.* 工匠
A skilled craftsman made this wooden chair.
這把木頭椅子是由一位技藝高超的工匠製作的。

❹ **afford** [əˈfɔrd] *vt.* 買得起；負擔得起
Dave can't afford to buy a new car right now.
戴夫目前買不起新車。

❺ **discount** [ˈdɪskaʊnt] *n.* 折扣 & [dɪsˈkaʊnt] *vt.* 打折
The store is having a sale, with discounts on all summer clothing.
這間商店正在進行特價活動，所有夏季服裝都有折扣。

❻ **guarantee** [ˌgærənˈti] *vt.* 保證，擔保 & *n.* 保證
The company guarantees the quality of its products.
這間公司保證其產品的品質。

❼ **Take it or leave it.**　接不接受隨便你。
I've already lowered the price as much as I can. Take it or leave it.
我已經盡可能降價了。買不買隨你。

經典旅遊實用句

Can you give me a [折扣]?　能不能再多給點 [折扣]？

用途：進一步要求更多折扣。

能不能再多打點折？
Can you give me a bigger discount?
　　　　　　　better price　更優惠的價格
　　　　　　　better deal　更划算的優惠

旅人私房筆記

各國著名紀念品小指南

日本	泰國	義大利
folding fan 摺扇	**Thai silk** 泰絲	**Murano glass** 穆拉諾玻璃

法國	澳洲	英國
macaron 馬卡龍	**kangaroo plush** 袋鼠玩偶	**red telephone box model** 紅色電話亭模型

Try It Out! 試試看
請依句意在空格內填入適當的字詞

某些市場可以接受講價，但百貨公司不接受。
It's acceptable to _____ in some markets, but not in department stores.

Ans｜ bargain

旅遊英語輕鬆聊
參考答案請掃描使用說明頁上 QR code 下載

Have you ever bargained for something while traveling? What happened?

CH 07 Shopping・購物

Unit 55
Requesting a Return or Refund
要求退貨退款

實境對話 GO！

B Bruce（布魯斯）　　**A** Amanda（亞曼達）

Amanda is speaking to Bruce, a customer service representative, about an item she wishes to return.

B Hi, there! How can I help you?

A Hello. I would like to return this dress, please.

B Is there something wrong with it?

A No, not really. It was a gift, but it's too small for me.

B Do you have the original receipt?

A Yes, my friend included the receipt with the gift. Here you are.

B Thank you.

A I'd like a refund, please, if that's possible.

B Unfortunately, this dress was on sale when it was purchased, so we can only offer to exchange it for something else of the same value.

A Oh, I wasn't aware of that.

B It's our store's policy, I'm afraid. Would you like to try on a bigger size?

A Um… not really.

B In that case, you can take a look around the store to see if there's anything else you would like. **Alternatively**, I can give you a store credit and you can spend it some other time.

A You really can't just give me the cash?

B I'm sorry, miss, but like I said: it's store policy.

A OK, I guess I'll take the store credit, then.

B Very well. Here's a gift card for the price of the dress: $29.95.

A Thank you.

> 亞曼達跟客服布魯斯說她要退貨一件商品。

B 您好！請問需要服務嗎？

A 你好。我要退這件洋裝。

B 請問是有什麼問題嗎？

A 不算是。這是人家送的，但我穿太小了。

B 請問您有原始收據嗎？

A 有，我朋友送的時候連收據一起給我了。在這裡。

B 謝謝您。

A 如果可以的話，我想要退錢。

B 不好意思，這件洋裝當初購買時是特價商品，所以我們只能讓您換購其他等值商品。

A 喔，我不知道有這種事。

B 抱歉，這是店內規定。您想試穿大一號看看嗎？

A 嗯……沒有很想耶。

B 那樣的話，您可以在店裡看看有沒有其他您喜歡的。不然，我可以給您一張禮券，您可以之後再用。

A 真的不能直接退現金給我喔？

B 很抱歉，小姐，但我說過了，店內規定就是這樣。

A 好吧，那我就拿禮券。

B 好的。這張是洋裝等值金額的禮券：29.95 美元。

A 謝謝。

旅遊字詞補給包

❶ **request** [rɪˈkwɛst] *vt. & n.* 請求；要求
Charlie politely requested another cup of coffee.
查理禮貌地要了另一杯咖啡。

❷ **refund** [ˈriˌfʌnd] *n.* 退貨還款 & [rɪˈfʌnd] *vt.* 退還
Bonnie was not satisfied with the item, so she requested a refund from the store.
邦妮對於商品不滿意，所以要求店家退款。

❸ **be aware of...**　　知道……；察覺到……
You must be aware of your surroundings to ensure your safety.
你必須對周遭環境有警覺心，以確保自身安全。

❹ **policy** [ˈpɑləsɪ] *n.* 規定，政策
The school's dress policy requires students to wear uniforms.
學校的服裝規定是要求學生穿制服。

❺ **alternatively** [ɔlˈtɝnətɪvlɪ] *adv.* 或者，要不
You can take the MRT to get there; alternatively, you can choose to take the bus.
你可以搭捷運去那裡，或者也可以選擇搭公車。

經典旅遊實用句

I would like to return [商品], please.　　我想要退……。

用途：表達退貨意圖。

我想要退這件洋裝。
I would like to return this dress, please.
　　　　　　　　　　these shoes　　這雙鞋
　　　　　　　　　　this item　　　這件商品
　　　　　　　　　　these boots　　這雙靴子

旅人私房筆記

「退貨理由」常見英文表達

原因	英文表達方式
尺寸不合	It doesn't fit me.
顏色不喜歡	I don't like the color.
自己已經有了	I already have one.
品項錯誤或不如預期	This isn't what I expected.
商品有瑕疵	The item has a defect.

Try It Out! 試試看

請依句意在空格內填入適當的字詞

遊客到國外玩時必須注意當地的習俗和傳統。

Tourists must _____ _____ _____ the local customs and traditions when visiting a foreign country.

Ans be aware of

旅遊英語輕鬆聊

參考答案請掃描使用說明頁上 QR code 下載

Imagine you can't get a refund. Would you prefer exchanging the item or getting store credit? Why?

CH 07 Shopping・購物

Unit 56 Duty-Free Shopping at the Airport

免稅店購物樂

實境對話 GO！

C Clerk（店員）　　**S** Sandra（珊卓）

While waiting for her connecting flight at Zurich Airport, Sandra is browsing in a duty-free shop.

C Hello, ma'am. Is there anything I can help you with today?

S I'm just checking out some of these perfumes.

C Oh yes, we have some of the top brands from European designers.

S I know. I'm blown away by how inexpensive they are.

C Well, if you buy them here at the airport's duty-free store, then you can avoid paying the import duty.

S This one has an amazing scent. I'll take a bottle.

C Certainly! It looks like this is the last bottle we have of this fragrance.

S Oh nice. I guess it's my lucky day.

C Let's head to the register. May I see your passport?

S Here you go.

C All right, would you like to add any other items? If you spend over $300 euros, then we can give you an additional 5% discount on your purchase.

S Really? Well, I guess I could add a bottle of red wine and a few boxes of these chocolates. They look delicious.

C Excellent choice. Those have been really popular this season. Would you like these items **wrapped** as gifts?

S No, that won't be necessary. Here's my credit card.

C Thank you very much and have a wonderful day!

珊卓於蘇黎世機場等待轉機的期間，在免稅商店裡四處閒逛。

C 這位女士，您好。有什麼我可以為您服務之處嗎？

S 我只是在看看這些香水而已。

C 原來如此，我們這邊有一些歐洲知名設計品牌的香水。

S 我知道。我很訝異它們竟然這麼便宜。

C 嗯，如果您在機場的免稅店裡購買這些商品的話，就毋須支付進口關稅。

S 這款的香味太迷人了。我要買一瓶。

C 沒問題！看來這是這款香水在本店的最後一瓶。

S 讚喔。我想今天是我的幸運日。

C 那我們就去收銀機那邊。我可以看一下您的護照嗎？

S 在這裡。

C 好的，那麼您還想加購其他商品嗎？如果您的金額超過三百歐元，我們就可以在本次消費為您多打 5% 的折扣。

S 真的嗎？那我想我可以多買一瓶紅酒，還有再買幾盒這種巧克力。它們看起來很好吃的樣子。

C 您真有眼光。那些都是本季的熱銷品項。這些商品要幫您做禮品包裝嗎？

S 不用，沒這個必要。我的信用卡在這。

C 非常感謝您的消費，祝您有個愉快的一天！

235

旅遊字詞補給包

❶ duty-free [ˌdjutɪˈfri] *a.* 免稅的
We bought some souvenirs at the duty-free shop in the airport before our flight.
我們在飛機起飛前到機場的免稅店買了些紀念品。

❷ blow sb away　　使某人大為驚訝 / 非常高興
The fact that Jonathan got up early and made everyone breakfast blew his mother away.
強納森早起幫大家做早餐這點讓他媽媽大感驚喜。

❸ inexpensive [ˌɪnɪkˈspɛnsɪv] *a.* 價格低廉的
This night market offers all kinds of inexpensive clothes and items.
這個夜市裡販售著各式各樣物美價廉的衣服和商品。

❹ scent [sɛnt] *n.* 香味；氣味
The room is filled with a lovely scent from those flowers.
房間裡瀰漫著那些花朵散發出的迷人香氣。

❺ fragrance [ˈfregrəns] *n.* 香水；芳香，香氣
Carrie enjoys trying different fragrances to find the perfect scent that matches her personality.
凱莉喜歡嘗試不同的香水，以找到最符合她個性的香味。

❻ (cash) register [(ˌkæʃ) ˈrɛdʒɪstɚ] *n.* 收銀機
The clerk rang up my groceries at the cash register.
店員把我買的雜貨輸入到收銀機上。

> ring up... / ring...up
> 把……輸入收銀機

❼ additional [əˈdɪʃənl̩] *a.* 額外的
We need additional workers to process this huge amount of data.
我們需要額外的人員來處理如此大量的數據資料。

❽ wrap [ræp] *vt.* 包，包裹　　三態為：wrap, wrapped, wrapped
It was so cold last night that Peggy wrapped herself in a blanket while watching TV.
昨晚太冷了，以致於佩姬在看電視時用毯子把自己裹住。

經典旅遊實用句

I'm just checking out some of these [商品].
我只是在看看這些……。

用途：客人回應自己只是在看看某些商品，常用於商店購物。

我只是在看看這些香水。
I'm just checking out some of these perfumes.
　　　　　　　　　　　　　　　　　watches　　手錶
　　　　　　　　　　　　　　　　　sunglasses　墨鏡

旅人私房筆記

免稅不手軟，瑞士好物帶著走！

Swiss Chocolate 瑞士巧克力

Swiss Watches 瑞士錶

Swiss Cheese 瑞士起司

Herbal Candies / Lozenges 瑞士草本潤喉糖

Swiss Army Knife 瑞士刀

Try It Out! 試試看　請依句意圈出適當的字詞

Compared to the other houses on this block, mine is rather (inexpensive / duty-free / additional).

Ans inexpensive

旅遊英語輕鬆聊　參考答案請掃描使用說明頁上 QR code 下載

Have you ever bought something at a duty-free store just because it felt like a "good deal"?

CH 07　Shopping・購物

237

Unit 57 Getting a Tax Refund
申請退稅

實境對話 GO !

- **O** Olivia（奧莉薇雅）
- **S** Staff member（工作人員）

Olivia walks up to a staff member at a duty-free store at the airport to ask some questions.

O Excuse me, do you know how I can get a tax refund for some items I bought during my travels?

S Of course. Do you have your receipts and passport with you?

O Yes, I have them here.

S Great. The tax refund **process** is quite simple. You can **proceed** to the tax refund counter just before the security check. They'll check your receipts and passport and then process your tax refund. You have to spend a **minimum** amount of US$50 for the refund, though.

O That sounds easy. Do I need to show them the products as well?

S No, but make sure they're in your carry-on luggage as you'll need to present them at the security check. Also, please note that tax refunds are only **applicable** for goods that are meant for personal use and not **commercial** purposes.

O Thanks for letting me know. I have been wondering about this for the whole trip.

S You're welcome. One more thing, make sure you get your tax refund processed before you've gone through security because you won't be able to do it at the gates.

O OK. I'll head directly to the tax refund counter now. Thank you again for your help.

S No problem. If you have any further questions, please feel free to come back anytime.

> 奧莉薇雅走向機場免稅店的工作人員以詢問一些問題。

O 不好意思，請問您知道我在旅行中購買的物品該如何退稅嗎？

S 當然。您身上有收據和護照嗎？

O 是的，我這裡有。

S 太好了。退稅流程非常簡單。您可以在安檢前前往退稅櫃檯辦理。他們會檢查您的收據和護照，然後處理您的退稅。不過，您必須花費至少五十美元才能退稅。

O 這聽起來很簡單。我是否也需要給他們看商品呢？

S 不用，但請確保它們在您的隨身行李中，因為您需要在安檢時出示它們。另外請注意，退稅僅適用於個人使用而非商業目的的商品。

O 謝謝您告訴我。整個旅程我都在想這個問題。

S 不客氣。還有一件事，請確保在通過安檢之前處理好退稅，因為您無法在登機口辦理。

O 好的。我現在直接去退稅櫃檯。再次感謝您的幫助。

S 不客氣。如果您還有任何問題，歡迎隨時回來。

旅遊字詞補給包

① **process** [ˈprasɛs] *n.* 過程 & *vt.* 處理
Getting fit again is a long and slow process.
再度恢復往日身材是個漫長的過程。
It will take seven days for your passport to be processed.
處理你的護照需要七天的時間。

❷ **proceed** [prəˈsid] *vi.* 繼續進行
proceed to V　（做完某事後）接下來做……
Judy sipped tea and proceeded to do some stretches.
茱蒂喝了口茶並開始做伸展運動。

❸ **minimum** [ˈmɪnɪməm] *a.* 最少的
The minimum age for entering nightclubs is 18.
進入夜店的最低年齡限制是十八歲。

❹ **applicable** [ˈæplɪkəbḷ] *a.* 適用的
Please check if the special offer is still applicable before you make a purchase.
在您購買之前，請確認這項特別優惠是否仍然適用。

❺ **commercial** [kəˈmɝʃəl] *a.* 商業的
Our company has enjoyed great commercial growth over the past five years.
過去五年來，我們公司在業績上有亮眼的成長。

❻ **feel free to V**　不要猶豫做……
If you have any problems, feel free to let us know.
如果有任何問題，儘管讓我們知道。

經典旅遊實用句

You can proceed to [地點].　你可以前往……辦理。

用途：指引旅客前往某處辦理。

您可以前往退稅櫃檯辦理。
You can proceed to the tax refund counter.
　　　　　　　　　the immigration area　入境審查區
　　　　　　　　　Gate B1　　B1 登機門

240

旅人私房筆記

好想退稅卻說不出口嗎？這些退稅英文小短句必須筆記！

出國購物好痛快，若是看到店家寫 Duty Free 表示免稅商店，通常消費滿一定額度並出示護照，即能符合免稅資格。另外看到 Tax Free 或是 Tax Refund 的店家，代表可以退稅。但是該怎麼跟店家詢問有關退稅的問題呢？

- **Does the price include tax?**
  這個價格含稅嗎？

- **Is this item tax free?**
  這個免稅嗎？

- **May I have a tax refund form, please?**
  請給我一張退稅單好嗎？

- **I want to claim a tax refund.**
  我想要申請退稅。

Try It Out! 試試看
請依句意在空格內填入適當的字詞

① 開班所需的最少學生人數為二十人。
The _____ number of students that is needed to open a class is 20.

② 這部電影是否會成為商業賣座片仍不確定。
Whether the movie will be a _____ success remains uncertain.

Ans ① minimum　② commercial

旅遊英語輕鬆聊
參考答案請掃描使用說明頁上 QR code 下載

What kind of items would you try to get a tax refund for when traveling?

CH 07　Shopping．購物

241

Notes

Chapter 08

Emergency
緊急狀況

Unit 58
Making an Appointment
掛號

Unit 59
Talking with the Doctor
向醫生說明病情

Unit 60
A Quick Stop at the Pharmacy
去藥局領藥

Unit 61
Reporting a Crime
警察局報案

Unit 62
Checking the Lost and Found
失物招領處

Unit 63
Asking for Directions
問路

Unit 64
Looking for a Restroom
找廁所

Unit 65
Losing Your Passport
遺失護照

243

Unit 58 Making an Appointment
掛號

實境對話 GO！

B Basil（貝索）　　**J** Jennifer（珍妮佛）　　**R** Receptionist（櫃檯小姐）

The following morning, Jennifer feels even worse, so she calls her local clinic to set up an appointment to see the doctor.

B Well, you aren't dizzy anymore, but I still think you need to see the doctor.

J Fine, I'll make an appointment. Can you get my phone for me?

B Here you are.

(Jennifer calls her clinic.)

J Hello, I'd like to make an appointment for today if possible.

R We have one opening at four o'clock. Does that work for you?

J That's fine. Is it Dr. Kashino?

R Sorry, she's out today. Dr. Hamilton is the only one available. If you would like to see Dr. Kashino, she'll be back tomorrow.

J No, that's fine. I think I need to be seen today.

R Can you tell me the reason for your visit?

J My throat is killing me, I'm dizzy, and I fainted yesterday. I have no idea what's going on!

244

R Yes, I think it's best you come in today. Can I get your **insurance** info, please?

J Yeah. I'm on Blue Cross, **account** number 86-75-309.

R Jennifer Lee, Blue Cross 86-75-309. Is that correct?

J Yes.

R We'll see you this afternoon, Ms. Lee.

隔天早晨，珍妮佛感到更加不舒服，所以她打給當地的診所預約一個門診掛號看醫生。

B 嗯，妳已經不會頭暈了，但我認為妳還是得看個醫生。

J 好吧，我會預約個門診。你可以幫我拿一下手機嗎？

B 拿去。

（珍妮佛打給診所。）

J 您好，如果可以的話我想預約今天的門診。

R 我們四點有個空檔。請問這個時段您方便嗎？

J 可以的。請問是樫野醫師看診嗎？

R 抱歉，她今日外出不在。漢密爾頓醫師是唯一的看診醫師。如果您想要看樫野醫師的門診，她明天會回來。

J 不用了，沒關係。我想我今天就需要看診。

R 能否請問您前來看診的原因？

J 我的喉嚨痛得受不了，我頭很暈，而且昨天還昏倒了。我完全不知道是怎麼回事！

R 是的，我認為您最好是今天就來看診。我可以跟您要您的保險資料嗎？

J 好的。我是在藍十字保險公司投保，帳號是 86-75-309。

R 珍妮佛·李，藍十字保險公司，帳號是 86-75-309。以上正確嗎？

J 沒錯。

R 那我們下午見了，李小姐。

旅遊字詞補給包

❶ **appointment** [əˋpɔɪntmənt] *n.* 預約；(正式的) 約會；會面；委任
I have an appointment with the doctor this afternoon.
我今天下午和醫生有約診。

❷ **clinic** [ˋklɪnɪk] *n.* 診所，診間
The rich man founded a free clinic.
這位有錢人創立了一間免費的診所。

❸ **if possible**　　如果可能的話
If possible, it's better to work in natural light.
如果可能的話，最好在天然光線下工作。

❹ **faint** [fent] *vi.* 暈厥；昏過去
When I saw my favorite singer in person, I nearly fainted.
我親眼看到最喜愛的歌手時幾乎快要暈倒。

❺ **insurance** [ɪnˋʃʊrəns] *n.* 保險；保險業
He filed a claim with his home insurance after the storm caused damage.
在暴風雨造成損壞後，他提出房屋險的理賠申請。

❻ **account** [əˋkaʊnt] *n.* 帳戶；帳目
Ella opened a new savings account for her children's education.
艾拉為孩子們的教育費開了一個新的儲蓄帳戶。

經典旅遊實用句

I'd like to make an appointment for [時間] if possible.
如果可以的話我想預約……的門診。

用途：表達希望盡快預約看診。

如果可以的話我想預約今天的門診。
I'd like to make an appointment for today if possible.

tomorrow morning	明天早上
this week	這個禮拜
this afternoon	今天下午

旅人私房筆記

不舒服怎麼說？出國不舒服也能開口求助！

旅行時若有身體不適，最怕的就是看醫生時不知道該怎麼表達，我們來看看幾個常見病徵的英文說法：

- **I have a sore throat.**
  我喉嚨痛。

- **I feel dizzy.**
  我覺得頭暈。

- **I have an upset stomach.**
  我胃不舒服。

- **I have a fever.**
  我發燒了。

- **I feel nauseous** [ˈnɔʃɪəs].
  我覺得想吐。

- **I have a runny nose.**
  我流鼻水。

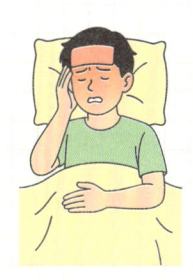

Try It Out! 試試看　請依句意在空格內填入適當的字詞

① 如果可以的話，事先訂好你的火車票。
　_____ _____, order your train ticket in advance.

② 賴瑞餓到差點昏倒。
　Larry was so hungry that he almost _____.

Ans　① If possible　② fainted

旅遊英語輕鬆聊　參考答案請掃描使用說明頁上 QR code 下載

Have you ever had an urgent need to see a doctor while traveling? What happened?

CH 08　Emergency・緊急狀況

247

Unit 59 Talking with the Doctor
向醫生說明病情

實境對話 GO！

D Doctor（醫生）　　**J** Jennifer（珍妮佛）

Jennifer describes her symptoms to the doctor.

D Hello, Ms. Lee. How are you feeling today?

J Not great.

D Can you describe your sickness?

J Yesterday, I suddenly felt ill. I got quite dizzy and fainted. Today, I've been coughing, and I've got this terrible sore throat.

D OK. Well, let me take a look at your throat and listen to your breathing. Open your mouth and say "ah."

J Ah!

D The back of your throat looks a little red and swollen. Now, take a deep breath.

(Jennifer takes a few deep breaths.)

D I think your lungs have some fluid in there.

J Do you have any idea what's wrong with me?

D OK. I think you have two illnesses. First, you've got a classic case of **bronchitis**. We should be able to treat that fairly quickly. But for the dizziness, we'll need to run some tests. Can I take a blood sample right now?

J Sure.

D When I receive the results, I will get back to you. For now, I can give you something for the bronchitis.

> 珍妮佛向醫生說明她的症狀。

D 您好,李小姐。您今天感覺如何?

J 不是很好。

D 您可以描述一下怎麼不舒服嗎?

J 昨天我突然感到不舒服。我頭很暈而且還昏倒了。今天我開始一直咳嗽,而且我的喉嚨非常地痛。

D 好的。嗯,請讓我看一下您的喉嚨,然後聽一下您的呼吸。請張開嘴巴然後說「啊」。

J 啊!

D 您的喉嚨後方看起來有些紅腫。現在請深呼吸。

(珍妮佛深呼吸了幾次。)

D 我認為您的肺部裡有些液體。

J 請問您知道我生了什麼病嗎?

D 好的,我認為您有兩個病症。首先,您得了典型的支氣管炎。我們應該很快就能進行治療。但對於頭暈的部分,我們需要做一些檢驗。我現在能幫您抽個血嗎?

J 當然。

D 等我收到檢驗報告,就會通知您。現在的話,我可以先給您一些治療支氣管炎的處方。

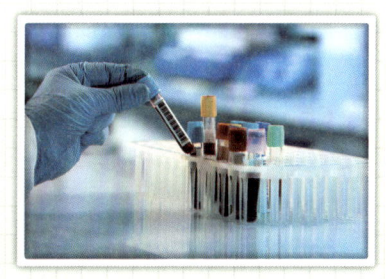

> 旅遊字詞補給包

❶ **symptom** [ˈsɪmptəm] *n.* 症狀；徵兆
A fever is a common symptom of the flu.
發燒是流感常見的症狀。

❷ **cough** [kɔf] *vi.* 咳；咳嗽
Please remember to cover your mouth when you cough.
咳嗽時請記得摀住嘴巴。

❸ **sore throat** [ˌsɔr ˈθrot] *n.* 喉嚨痛；喉嚨不適
Because Tim caught a cold a few days ago, he has a runny nose and sore throat.
因為提姆前幾天感冒了，所以他流鼻涕又喉嚨痛。

❹ **breathing** [ˈbriðɪŋ] *n.* 呼吸；微風
I've written you a prescription for your difficulty breathing.
我為你開了治療呼吸困難的處方。

❺ **fluid** [ˈfluɪd] *n.* 液體；流體，流質
The nurse told him to drink plenty of fluids.
護理師告訴他要多攝取流質。

❻ **bronchitis** [brɑŋˈkaɪtɪs] *n.* 支氣管炎
Dennis developed bronchitis after a bad cold.
丹尼斯在重感冒後得了支氣管炎。

> 經典旅遊實用句

Can you describe your [症狀/狀況]?
你可以描述一下你的……嗎？

用途：詢問病人具體症狀。

您可以描述一下怎麼不舒服嗎？
Can you describe your sickness?

pain	疼痛
symptoms	症狀
discomfort	不適

旅人私房筆記

旅途中看醫生：症狀、檢查一次搞懂！

出國就醫需要做一些相關檢查，可別搞不清楚這些英文是什麼呀！

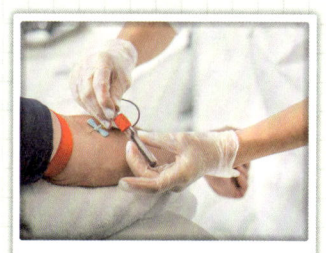

blood test / take a blood sample
驗血

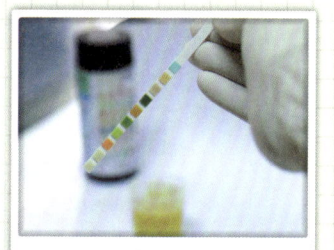

urine [ˈjurɪn] test
驗尿

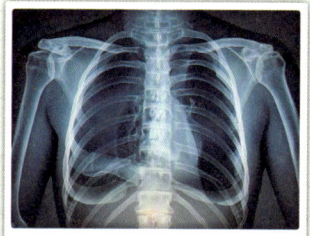

X-ray
照 X 光

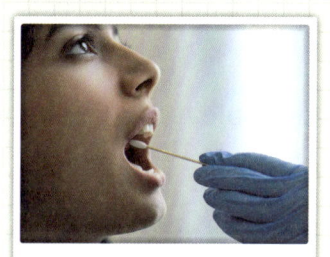

throat swab [swɑb] test
咽喉拭子檢查

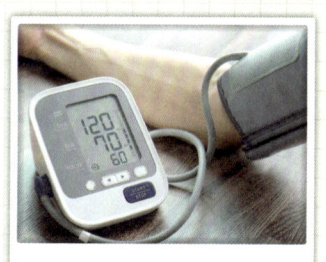

check your blood pressure
量血壓

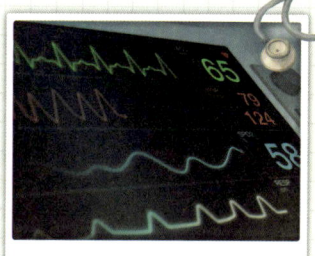

ECG / EKG (electrocardiogram [ɪˌlɛktroˈkɑrdɪəˌgræm])
心電圖檢查

Try It Out! 試試看　請依句意選出適當的字詞

I'm not feeling well because I have been _____ all day.

Ⓐ coughing　Ⓑ smiling　Ⓒ dreaming　Ⓓ holding

Ans Ⓐ

旅遊英語輕鬆聊　參考答案請掃描使用說明頁上 QR code 下載

Would you rather take medicine or get an injection? Why?

CH 08　Emergency・緊急狀況

251

Unit 60 A Quick Stop at the Pharmacy
去藥局領藥

實境對話 GO!

P Pharmacist（藥師）　　**J** Jennifer（珍妮佛）

Jennifer stops by the drugstore to pick up her medicine on her way home.

P May I have your insurance card and prescription, please?

J Here you are.

P Ms. Lee. We've got a few things for you here. This red one is just a painkiller. Take two every four hours, but only when you need them. Make sure not to drink any alcohol with the painkillers.

J What about tea or coffee? Can I have caffeine?

P That shouldn't be a problem. These purple ones you take just before bed. They'll help you fall asleep and stop your cough.

J What about these giant green capsules?

P Those are the antibiotics for your bronchitis. Take one with breakfast and one after dinner. If you aren't feeling better in a few days, you should go back to Dr. Hamilton.

J So purple before bed, green twice a day, and red as I need it?

P That's right. Here you are. Any other questions?

J Nope, I think I've got it. Can I take cough syrup, too?

P Most cough medicines have alcohol in them, so I wouldn't suggest it.

珍妮佛在回家途中經過藥局領藥。

P 請問能跟您要您的健保卡以及處方箋嗎？

J 在這裡。

P 李小姐。這裡有些東西要給您。這個紅色的是止痛藥。每隔四小時吃兩顆，但只在您需要的時侯才服用。服用這個止痛藥時請不要喝酒。

J 那請問茶或是咖啡呢？我能攝取咖啡因嗎？

P 應該沒問題。這些紫色的要睡前服用。它們會幫助您入眠及止咳。

J 那這些大顆的綠色膠囊呢？

P 那些是治療您的支氣管炎的抗生素。請於早餐時服用一顆，晚餐後服用一顆。如果您幾天後仍未好轉，那您應該回去找漢密爾頓醫師。

J 所以睡前吃紫色的，綠色的一天吃兩次，然後紅色的是有需要才吃？

P 沒錯。您的藥在這邊。還有其他問題嗎？

J 沒有，我想我了解了。請問我可以也喝咳嗽糖漿嗎？

P 大多數的咳嗽糖漿含有酒精成分，因此我不建議這樣做。

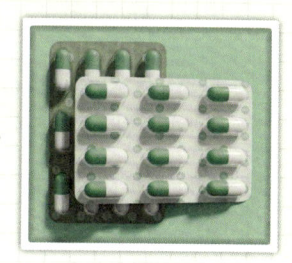

旅遊字詞補給包

❶ pharmacy [ˈfɑrməsɪ] *n.* 藥局；配藥學
I need to go to the pharmacy to pick up my prescription.
我需要去藥局領我的處方藥。

❷ pharmacist [ˈfɑrməsɪst] *n.* 藥劑師
The pharmacist explained how to take the medication properly.
那位藥劑師解釋了如何正確服用藥物。

❸ pick up 拿取；撿起
Penny has a package that she needs to pick up on her way to work.
潘妮有個包裹需要她在上班途中去拿。

❹ **prescription** [prɪˋskrɪpʃən] *n.* 處方，藥方
You need a doctor's prescription to buy this strong pain medication.　你需要醫師處方才能買這種強效止痛藥。

❺ **caffeine** [kæˋfin] *n.* 咖啡因
Chloe tries to avoid caffeine in the evening so she can sleep better.
克洛伊試著避免在晚上攝取咖啡因，這樣她才能睡得更好。

❻ **fall asleep**　入睡；進入睡眠狀態
The more you worry, the harder it is to fall asleep.
你越煩惱就越難睡著。

❼ **antibiotic** [ˌæntɪbaɪˋɑtɪk] *n.* 抗生素，抗菌素
It's important to finish the entire course of antibiotics even if you feel better.　即使你感覺好多了，也要吃完整個療程的抗生素藥品。

❽ **cough syrup** [ˋkɔf ˌsɪrəp] *n.* 咳嗽糖漿，止咳藥水
You can buy cough syrup at the pharmacy without a prescription.
你可以在藥局購買非處方止咳糖漿。

經典旅遊實用句

Make sure not to [動作] with this medication.
服用此藥時請不要……。

用途：提醒服藥時避免的行為或飲食。

服用這個止痛藥時請不要喝酒。
Make sure not to <u>drink any alcohol</u> with the painkillers.
　　　　　　　　<u>drive</u>　　開車

旅人私房筆記　　　藥品類型的說法

本文中的 capsule [ˋkæps!̩] 這個名詞表「膠囊」。另外一些常見的藥品類型還有 pill（藥丸）、tablet（藥片）、drops（滴劑）等。像 sleeping pill 就是「安眠藥」。例：

| **capsule** 膠囊 | This bottle of vitamin C contains 200 <u>capsules</u>.
這瓶維他命 C 含有兩百粒膠囊。 |

tablet　藥片		The doctor gave me some tablets to help control my frequent headaches. 醫師給了我一些藥片幫助我控制頻繁的頭痛。
pill　藥丸		Judy has to rely on sleeping pills in order to fall asleep. 茱蒂要靠安眠藥才能睡著。
drop　滴劑		The doctor prescribed drops for my ear infection. 醫生開了滴劑來治療我的耳朵感染。
powder　藥粉		The ancient medicine was made from a blend of various herbal powders. 這種古老的藥物是由多種草藥粉末混合製成的。
syrup　糖漿		If your symptoms don't improve after taking the syrup for a week, consult your doctor. 如果您服用糖漿一週後症狀沒有改善，請諮詢您的醫生。

Try It Out! 試試看　請依句意在空格內填入適當的字詞

❶ 有些人對咖啡因的作用非常敏感。
　Some people are very sensitive to the effects of _____.

❷ 這個止咳糖漿嚐起來很甜。
　This _____ _____ tastes quite sweet.

❸ 你可以向藥劑師詢問怎麼選止痛藥。
　You can ask the _____ for advice on choosing a pain reliever.

Ans ❶ caffeine　❷ cough syrup　❸ pharmacist

旅遊英語輕鬆聊　參考答案請掃描使用說明頁上 QR code 下載

Have you ever gotten confused about when or how to take your medicine?

Unit 61 Reporting a Crime
警察局報案

實境對話 GO！

P Police officer（警察）　　**K** Kevin（凱文）

Kevin is at the police station reporting a robbery.

P Hello sir, you said you would like to make a report?

K Yes. I was robbed about an hour ago.

P Can you tell me where it happened?

K Uhhh...downtown in an alley. By the movie theater. I don't remember the name of the street.

P Was it a large movie theater?

K Yeah, I was on my way home from a movie.

P OK, so probably the Golden Theater. Can you describe the robber?

K He was fairly tall, skinny, with long, brown hair. I think he had a gun, but I couldn't tell if it was real.

P Did he have a beard?

K Nope, no beard. He was acting a bit nervous, so when he pulled out the gun, I did what he asked.

P And what did he ask for?

K Just my wallet. But that has everything I need! I just lost my passport last week.

P Relax, your description fits someone we've been tracking. There's a good chance we can get it back.

K That's a relief to hear. Thank you.

凱文在警局報一件搶案。

P 先生你好，你說你要報案嗎？

K 是的。我大約一小時前被搶。

P 你能告訴我案發地點嗎？

K 呃……在市中心的一條小巷子裡。在電影院旁邊。我不記得街名了。

P 是大間的電影院嗎？

K 對，我當時看完電影正要回家。

P 好的，所以很有可能是黃金戲院。你能形容一下那個搶匪嗎？

K 他相當地高瘦，留著棕色長髮。我想他有一把槍，不過我看不出那是不是真的。

P 他有留鬍子嗎？

K 沒有，他沒有鬍子。他表現得有點緊張，所以他拔出槍時說什麼我都照做了。

P 那他向你要了什麼？

K 就只有我的錢包而已。但是我需要的所有東西都在錢包裡面！我上週才剛遺失護照。

P 放心，你的描述符合我們正在追緝的一名嫌犯。我們很有機會能找回你的錢包。

K 聽你這麼說我就放心了。謝謝你。

旅遊字詞補給包

① **robbery** [ˈrɑbərɪ] n. 搶劫
rob [rɑb] vi. 搶劫 & vt. 搶奪
robber [ˈrɑbɚ] n. 強盜，搶匪

三態為：rob, robbed, robbed

The robbery led to a **shoot-out** between the robbers and the police.
這起搶案引發警匪槍戰。

shoot-out [ˈʃutˌaʊt] n. 槍戰

CH 08 Emergency・緊急狀況

257

The masked men tried to rob the bank.
那些蒙面男子試圖搶劫銀行。
Witnesses described the robber as a tall and thin man.
目擊者描述搶匪是高瘦的男子。

❷ **downtown** [ˌdaʊnˈtaʊn] *adv.* 在市區，往市區 & *a.* 市中心的
Ella works downtown in a large office building.
艾拉在市中心一棟大型辦公大樓裡工作。

❸ **alley** [ˈælɪ] *n.* 小巷
Dan used to take a shortcut to school through that alley.
丹以前都從那條小巷抄近路去學校。

❹ **beard** [bɪrd] *n.* 下巴及兩耳下方的鬍鬚
Jacob's beard was starting to turn gray.
雅各的鬍鬚開始變白。

❺ **wallet** [ˈwɑlɪt] *n.* 皮夾，錢包
Micky lost his wallet on the bus this morning.
米奇今天早上在公車上掉了皮夾。

❻ **track** [træk] *vt.* 追蹤；留下 (泥巴等) 痕跡於……
The police are tracking the suspects' getaway car.
警方正追蹤嫌疑犯使用的接應車輛。

經典旅遊實用句

I was on my way + [介詞] **+** [地點] **.**　　我當時是在去……的路上。

用途：警察詢問案發場景。
我當時在看完電影要回家的路上。
I was on my way **home from a movie.**
　　　　　　　 to the airport　　前往機場
　　　　　　　 to the hotel　　　去飯店

258

旅人私房筆記

嫌犯大追蹤！英語描述術大公開

在向警察報案時，清楚描述嫌犯外貌有助於加快調查速度。

身高體型

- **tall** （高）
- **short** （矮）
- **medium height** （中等身高）
- **slim** （瘦）
- **muscular** （壯碩）
- **chubby** （胖胖的）

髮型與臉部

- **bald** （禿頭）
- **long / short / curly / straight hair** （長 / 短 / 捲 / 直髮）
- **beard / mustache** （鬍子 / 小鬍子）
- **glasses** （眼鏡）

穿著

- **wearing a jacket / a hoodie / a cap / sunglasses**
  （穿著夾克 / 穿著帽T / 戴帽子 / 戴墨鏡）
- **in black / blue / dark clothes** （穿黑色 / 藍色 / 深色衣物）
- **carrying a backpack / bag** （背著包包）

Try It Out! 試試看
請依句意在空格內填入適當的字詞

你可以在線上追蹤你的包裹，查看它何時會送達。
You can _____ your package online to see when it will arrive.

Ans track

旅遊英語輕鬆聊
參考答案請掃描使用說明頁上 QR code 下載

Your wallet was stolen. What do you most hope to get back?

CH 08　Emergency・緊急狀況

Unit 62 Checking the Lost and Found
失物招領處

實境對話 GO！

J Julian（朱利安）　　**R** Ronda（朗達）

Julian is speaking to Ronda, a transportation official, at a train station.

J Excuse me. Is this the office that collects lost items?

R Yes, I'm in charge of the lost and found. How can I help you?

J I left my bag on a bench on one of the train platforms last night.

R Well, if someone turned it in, we should have it here.

J I hope that's what happened. I was on the train about 10 minutes before I realized it was missing.

R What does it look like?

J It's a small, brown messenger bag with a grey strap and two silver buckles.

R All right. Was there anything valuable inside?

J There was a nice pair of sunglasses inside. Apart from that, there were some important work documents in there.

R Give me a second.

(Ronda goes to search a storeroom behind her desk.)

R Is this the bag you lost?

J Yes! That's it. I'm so glad it wasn't stolen.

R Take a look and see if everything is inside.

J Oh, no. It looks like the sunglasses are gone. Oh, well. At least my work papers are still here.

R Sorry to hear that. Before you go, you need to fill out this form and sign it.

J No problem. Thanks for your help.

> 朱利安正在火車站與站務人員朗達交談。

J 不好意思。這裡是失物招領處嗎？

R 是的，我是失物招領處的負責人。有什麼需要幫忙的嗎？

J 我昨晚把包包忘在其中一個火車月臺的長椅上了。

R 嗯，如果有人拿過來，這裡應該會有。

J 希望如此。我在火車上大約十分鐘後才意識到它不見了。

R 它長什麼模樣？

J 它是一個小巧的咖啡色郵差包，配有灰色背帶和兩個銀色扣環。

R 好的。裡面有什麼貴重物品嗎？

J 裡面有一副漂亮的太陽眼鏡。除此之外，還有一些重要的工作文件。

R 等我一下喔。

（朗達到她辦公桌後面的貯藏室裡尋找。）

R 這個是你遺失的包包嗎？

J 是的！是這個沒錯。真高興它沒有被偷走。

R 看一看是不是所有東西都在裡面。

J 噢，不。看來太陽眼鏡不見了。喔，好吧。至少我的工作文件還在。

R 很遺憾聽到這個消息。在你走之前，你需要填寫這個表格並簽名。

J 沒問題。謝謝妳的幫忙。

旅遊字詞補給包

❶ lost and found　　失物招領處
Gina found a wallet in the museum and handed it in to the lost and found.
吉娜在博物館撿到一個皮夾，並把它交給了失物招領處。

❷ a messenger bag　　郵差包
Larry carries his books and laptop in a messenger bag.
賴瑞把書和筆記型電腦放在郵差包裡。

❸ strap [stræp] *n.* 背帶
The strap of Jessie's bag broke, and all her belongings fell out.
潔西包包的帶子斷了，裡面的所有東西都掉了出來。

❹ buckle [ˈbʌkḷ] *n.* 扣環
Randy had to buy a new belt because the buckle on his old one was broken.
藍迪舊皮帶上的扣環壞了，必須買條新的。

❺ storeroom [ˈstɔrˌrum] *n.* 貯藏室
We keep extra supplies in the storeroom.
我們把多餘的用品放在貯藏室裡。

經典旅遊實用句

I left my bag [介詞] [地點] last night.　　我昨晚把包包忘在⋯⋯。

用途：說明遺失物品的地點。

我昨晚把包包忘在其中一個火車月臺的長椅上了。
I left my bag on a bench on one of the train platforms last night.

　　　　　on the train　　在火車上
　　　　　in the waiting area　　在候車區

旅人私房筆記

Lost & Found 英語實用表達

在國外旅遊弄丟東西實在是很麻煩，到了失物招領處要怎麼應對表達，也是一門功夫，我們來看一些常見的英文說法。

中文	常見英文說法
失物招領處	Lost and Found (Office) / Lost Property Desk
認領物品	claim an item
辦理遺失登記	report a lost item
留下聯絡資訊	leave your contact information
詳細描述物品	give a detailed description of the item

Try It Out! 試試看　請依句意在空格內填入適當的字詞

如果你撿到任何貴重物品，請把它送到失物招領處。
If you find anything valuable, please bring it to the _____ _____ _____.

Ans　lost and found

旅遊英語輕鬆聊　參考答案請掃描使用說明頁上 QR code 下載

> If you lost your bag in a foreign country, how would you explain it to the local police?

Unit 63 Asking for Directions
問路

實境對話 GO！

V Victor（維克多）　　**K** Kim（小金）

> Victor is lost, so he approaches Kim for directions to the Museum of Natural History.

V Excuse me, miss. I was wondering if you know how to get to the Museum of Natural History. I seem to have taken a wrong turn somewhere and my phone just ran out of battery.

K The Museum of Natural History? That's close to the city center.

V Is that quite far from here?

K It's not very far, but you have wandered off course quite a bit.

V Oh, really? I thought I knew where I was going. I guess I'm really quite lost.

K Don't worry. You are still within walking distance. I'll look up the directions for you.

V That would be great. Thanks so much.

K Cross back over this bridge and keep going straight for two blocks. Turn right, then keep going until you see a large statue.

V The one of a big bear?

K Yes, that one. The Museum of Natural History is just across the street from it.

V All right. Over the bridge, straight, turn right, and then **continue** until I see the bear.

K You've got it!

V I hope I'll remember that!

K Don't worry. Once you get a little closer to the city center, there are many signs that will point you in the right direction.

V Thank you for helping me out. You're a real **lifesaver**.

K It's no problem at all. Enjoy the museum!

V Thanks!

> 維克多迷路了，所以走向小金詢問前往自然歷史博物館的路線。

V 不好意思，小姐。請問你知道怎麼去自然歷史博物館嗎？我好像在某個地方轉錯方向了，而且我的手機電池剛好沒電了。

K 自然歷史博物館？它離市中心很近。

V 離這兒還很遠嗎？

K 不會很遠，但是你已經有點太偏離路線了。

V 哦，真的嗎？我還以為我知道該往哪走。我想我真的是迷路了。

K 別擔心。你仍然在走路可到的距離之內。我來為你找路。

V 那太棒了。非常感謝。

K 往後穿過這座橋，繼續向前直走兩個街區。向右轉，然後一直走，直到看到一個大型雕像。

V 是那座大熊的雕像嗎？

K 對，就是那一個。自然歷史博物館就在它前方的街對面。

V 好的。過橋，直走，右轉，然後繼續走直到我看到熊。

K 沒有錯！

V 我希望我能記住這些！

K 別擔心。一旦你離市中心靠近一點，就會有許多標誌指引你朝著正確的方向前進。

V 謝謝妳幫助我。妳真的是個救星。

K 千萬別客氣。好好參觀博物館！

V 謝謝！

旅遊字詞補給包

❶ directions [dəˈrɛkʃənz] *n.* (行路的) 指引 (恆為複數)
A very helpful local gave Andy directions to the airport.
一位樂於助人的當地人為安迪指路去機場。

❷ battery [ˈbætərɪ] *n.* 電池
The red light means the battery is charging.
這顆紅燈發亮表示電池在充電。

❸ wander [ˈwɑndɚ] *vi.* 漫遊；閒逛
It isn't safe for you to be wandering around this neighborhood at night.
你晚上在這附近閒逛不安全。

❹ off course　　偏離方向
The boat went off course during the storm.
那艘船在暴風雨中偏離了航線。

❺ statue [ˈstætʃu] *n.* 雕像
The statue is one of the artist's best pieces.
這座雕像是該藝術家最棒的作品之一。

❻ continue [kənˈtɪnju] *vi.* (朝一方向) 繼續走
The hiker continued along the road for a while.
健行客繼續沿著這條路走了一段時間。

❼ lifesaver [ˈlaɪfˌsevɚ] *n.* 救星、幫忙解決困難的人或物
Sam offered me a ride home, which was a total lifesaver.
山姆載我回家，這真是幫了我大忙。

經典旅遊實用句

Excuse me, I was wondering if you know how to get to [地點]. 不好意思，請問你知道怎麼去……嗎？

用途：有禮貌地詢問怎麼走

不好意思。請問你知道怎麼去自然歷史博物館？
Excuse me. I was wondering if you know how to get to the Museum of Natural History.

the train station	火車站
the post office	郵局
the Central Park	中央公園

旅人私房筆記　　　　問路小字典

在旅途中問路，這裡教你輕鬆表達常見表示方向的用語。

- **be close to...**　接近 / 鄰近……
- **be far from...**　距離……很遠
- **cross (over) the bridge / road / street / river...**
  過橋 / 馬路 / 街 / 河……
- **go straight**　直走
- **turn right / left**　右 / 左轉

Try It Out! 試試看　　請依句意圈出適當的字詞

You might get lost if you (approve / observe / wander) into the woods by yourself.

　　　　　　　　　　　　　　　　Ans　wander

旅遊英語輕鬆聊　　參考答案請掃描使用說明頁上 QR code 下載

When do you think would be the worst time for your phone to die?

Unit 64 Looking for a Restroom
找廁所

實境對話 GO！

S Sid（席德）　　**B** Blanka（布蘭卡）

Sid is in a bar and needs to use the bathroom, so he asks a waitress, Blanka, for help.

S Excuse me. Where is your bathroom, please?

B I'm sorry, sir. It's closed for maintenance. But there are public restrooms you can use.

S Oh... Where can I find them? They're not too far away, are they?

B No, they're not far. When you exit the bar, turn left. They're about fifty meters down the road on your left-hand side, opposite the entrance to the beach. You'll see the signs.

S OK, thank you.

B Do you have any change with you? You'll need to pay one euro to use them.

S Oh, no, I don't. I just have my card with me.

B That's fine—you can tap your contactless card to enter.

S Do a lot of the public restrooms here require a payment?

B Yes, it's quite common in this country. The fee helps to pay for cleaning and maintenance and ensures a more hygienic experience.

S I see. Thanks for your help.

B It's my pleasure, and apologies again for the inconvenience. Our bathroom will be working again tomorrow.

S No worries. Could I trouble you to look after my bag while I'm away?

B Sure thing. I'll keep an eye on it.

S Thank you.

席德在酒吧裡想上洗手間，於是問女服務生布蘭卡。

S 不好意思，請問洗手間在哪裡？

B 抱歉，先生。我們的洗手間因維修暫停使用。不過您可以使用公廁。

S 喔……在哪裡？沒有很遠吧？

B 不遠。走出酒吧後左轉，沿路走個五十公尺左右，它會在左邊，對面是海灘的入口處。您會看到牌子。

S 好，謝謝。

B 您身上有零錢嗎？使用公廁要一歐元。

S 喔，沒有耶。我只有帶信用卡。

B 沒關係 —— 用感應式信用卡也進得去。

S 這裡的很多公廁都要付費嗎？

B 是的，在這個國家很普遍喔。付費有助於支付清潔和維修費用，確保使用時有更衛生的環境。

S 了解。謝謝幫忙。

B 不客氣，我們再次為造成您的不便道歉。廁所明天就可以恢復使用了。

S 沒關係。我去上廁所的時候，可以麻煩幫忙看一下我的包包嗎？

B 當然可以。我幫您看著。

S 謝謝。

旅遊字詞補給包

❶ left-hand [ˈlɛftˈhænd] *a.* 左邊的，左側的
Cars in Taiwan are left-hand drive.　臺灣的汽車是左駕。

❷ opposite [ˈɑpəzɪt] *prep.* 在……的對面 & *a.* 相對的；相反的 & *n.* 相反的情形 & *adv.* 在對面
There's a small café opposite the museum entrance.
博物館入口的對面有一間小咖啡館。

❸ ensure [ɪnˈʃʊr] *vt.* 確保
Please ensure that all the doors are locked before you leave the building.
請確認所有的門在你離開建築物之前都已經鎖上了。

❹ hygienic [ˌhaɪdʒɪˈɛnɪk] *a.* 衛生的
The restaurant received high marks for its hygienic food preparation practices.
這家餐廳因其衛生的食物製作程序而獲得高分。

❺ look after sb/sth　看管 / 照顧（某人 / 物）
Could you look after my cat while I'm on vacation?
我度假的時候你能幫我照顧我的貓嗎？

❻ keep an eye on sb/sth　留意；照看（某人 / 物）
Keep an eye on the trail to avoid getting lost while hiking.
健行時要看好路徑以免迷路。

經典旅遊實用句

It's closed for [關閉的原因].　它因……而關閉。

用途：說明某設施暫時關閉的原因
我們的洗手間因維修而暫停使用。
Our bathroom is closed for maintenance.
　　　　　　　　　　cleaning　　清潔
　　　　　　　　　　repairs　　　維修
　　　　　　　　　　inspection　 檢查

旅人私房筆記

說對才找得到：各國的「公共廁所」怎麼說？

英國 — **public toilet, gents / ladies, WC**
public toilet 是最直接的說法，公園、車站常見 "gents"（男）/ "ladies"（女）。

美國 — **public restroom, restroom**
"restroom" 聽起來比較禮貌，公共場所（如商場、機場）會用這個詞。

加拿大 — **washroom**（尤其是加拿大英語）
加拿大最常見用法；既禮貌又中性，公共或私人場所皆通用。

澳洲 — **public toilet, toilet block**（公園或露營區常見）
"toilet" 是常見說法，但會加上 "public" 以表明非私人空間。

國際通用 — **WC,**
"WC" 是歐洲多國共通用法，字面為 "Water Closet"。

Try It Out! 試試看　請依句意在空格內填入適當的字詞

服務檯位於購物中心主要入口的對面。
The information desk is located _____ the main entrance of the shopping mall.

Ans opposite

旅遊英語輕鬆聊　參考答案請掃描使用說明頁上 QR code 下載

How would you react if you were charged to use the restroom?

Unit 65 Losing Your Passport
遺失護照

實境對話 GO！

V Vernon（弗農）   **S** Sylvia（席維亞）

Vernon has lost his passport while on vacation, so he is at his country's embassy, speaking to Sylvia.

V Good morning. Could you help me? I've lost my passport. I think it might've been stolen.

S I'm sorry to hear that. Have you reported it to the police?

V Yes. They gave me a report—I have it here. Could you explain how I should apply for a replacement passport?

S We can't issue a full-validity passport right now, but we can give you an emergency one.

V What do I need to do?

S You need to complete this application form. We'll need the completed form, a new passport photo, and some form of photo ID if possible. Do you have a digital copy of your passport?

V Yes, I've got a photo of it on my phone.

S Great! That'll help speed things up.

V OK, thanks. How much is the fee?

S It's $165. You'll still need to apply for a full-validity passport when you return home, but you won't need to pay any extra.

V How long will it take to receive my emergency passport? I am due to return home in three days.

S We can issue one within 24 hours. You can't travel to a third country with it, but you can use it to return home.

V OK. Thank you.

S You're welcome.

弗農在度假時弄丟護照，所以他去了本國大使館，正與職員席維亞談話。

V 早。我需要幫忙。我的護照不見了。可能是被偷了。

S 很遺憾發生這種事。您報警了嗎？

V 有。他們給了我一份報案證明 —— 在這裡。可以說明一下怎麼申請補發護照嗎？

S 我們沒辦法當下就簽發正式護照，但可以先給您一本緊急護照。

V 怎麼辦理？

S 您需要填一下這份申請表。我們需要填寫完整的表格、一張新的護照用照片；如果有的話，請另外提供有照片的身分證件。您有護照的電子檔嗎？

V 有，我的手機裡有護照的照片。

S 太好了！這樣可以加快辦理速度。

V 好的，謝謝。費用是多少？

S 165 美元。您回國後仍需要申辦正式護照，但不需要再付費。

V 緊急護照多久可以拿到？我預計三天後回國。

S 我們可以在 24 小時內核發。您不能拿這本護照去其他國家，但可以回國。

V 好的。謝謝。

S 不客氣。

旅遊字詞補給包

❶ **embassy** [ˈɛmbəsɪ] *n.* 大使館
Willy needs to go to the embassy in person to process his visa extension.
威利需要親自前往大使館辦理簽證加簽。

❷ **replacement** [rɪˈplesmənt] *n.* 替代品；替代人選
I ordered a replacement part for my car online.
我在網路上訂購了我汽車的一個替換零件。

❸ **issue** [ˈɪʃu] *vt.* 頒發；發行 & *n.* 重大議題；爭議；(期刊) 一期
Banks usually issue ATM cards to new customers who open savings accounts.
銀行通常會核發提款卡給開設儲蓄帳戶的新客戶。

❹ **validity** [vəˈlɪdətɪ] *n.* (法律上的) 有效，合法性 (不可數)
We need to confirm the validity of this driver's license.
我們需要確認這張駕照的有效性。

❺ **emergency** [ɪˈmɝdʒənsɪ] *n.* 緊急事件，緊急狀況
The emergency services responded quickly to the accident.
緊急救難單位迅速趕到事故現場。

❻ **speed up** 提高速度，加速
The new equipment will speed up the manufacturing process.
這批新設備將加快製造流程。

經典旅遊實用句

Could you explain how I should [補發護照]? 請問我該怎麼……？

用途：禮貌詢問申請流程。
請問我該怎麼申請補發護照？
Could you explain how I should apply for a replacement passport?
　　　　　　　　　　　　　　 get a new passport　　拿到新護照
　　　　　　　　　　　　　　 renew my passport　　 換發護照

旅人私房筆記 — 認識大使館

Embassy 大使館

是一個國家在別的國家設立的辦公室，處理兩國外交事務並幫助自己國家的國民解決問題。

大使館有哪些人？

職稱	職務
ambassador （大使）	代表自己的國家
consul / consular officer （領事 / 領事官）	幫旅客補辦護照、處理緊急事件
visa officer （簽證官）	幫外國人審查簽證
security staff （安全人員）	保護安全
local staff （當地員工）	處理日常工作

Try It Out! 試試看
請依句意在空格內填入適當的字詞

我們需要加快生產流程以趕上交貨日期。
We need to _____ _____ the production process to meet the deadline.

Ans speed up

旅遊英語輕鬆聊
參考答案請掃描使用說明頁上 QR code 下載

What do you think embassies should offer to make people feel less stressed?

Notes

Notes

Notes

國家圖書館出版品預行編目（CIP）資料

旅遊英語這樣學最有效！：65個情境×真人示範×
互動口說練習 = Travel English made easy /
賴世雄作. -- 初版. -- 臺北市：常春藤數位出版
股份有限公司, 2025.08
面； 公分. -- (常春藤生活必讀系列；BA25)
ISBN 978-626-7225-98-1 (平裝)

1. CST：英語　2. CST：旅遊　3. CST：會話

805.188　　　　　　　　　　　　114009806

填讀者問卷
送熊贈點

常春藤生活必讀系列【BA25】
旅遊英語這樣學最有效！：65個情境×真人示範×互動口說練習

總　編　審	賴世雄
終　　　審	梁民康
執行編輯	許嘉華
編輯小組	畢安安・Nick Roden・Brian Foden
排版設計	王玥琦・王穎緁・林桂旭
封面設計	王穎緁
錄　　　音	林政偉
播音老師	Leah Zimmermann・Michael Tennant
法律顧問	北辰著作權事務所蕭雄淋律師
出　版　者	常春藤數位出版股份有限公司
地　　　址	臺北市忠孝西路一段33號5樓
電　　　話	(02) 2331-7600
傳　　　真	(02) 2381-0918
網　　　址	www.ivy.com.tw
電子信箱	service@ivy.com.tw
郵政劃撥	50463568
戶　　　名	常春藤數位出版股份有限公司
定　　　價	399元
出版日期	2025年8月　初版／一刷

©常春藤數位出版股份有限公司 (2025) All rights reserved.　　　　　　Y000081-3598
本書之封面、內文、編排等之著作財產權歸常春藤數位出版股份有限公司所有。未經本公司書面同意，
請勿翻印、轉載或為一切著作權法上利用行為，否則依法追究。

如有缺頁、裝訂錯誤或破損，請寄回本公司更換。　　　　　【版權所有　翻印必究】
服務時間：週一至週五 9：00～17：00（國定假日公休，若有其他異動，依官網公告為主）